A GUIDE TO WRITING AS AN ENGINEER

A GUIDE TO WRITING AS AN ENGINEER

David Beer
Department of Electrical and Computer Engineering
University of Texas at Austin

David McMurrey
IBM Corporation

JOHN WILEY & SONS, INC.
NEW YORK • CHICHESTER • BRISBANE • TORONTO • SINGAPORE

ACQUISITIONS EDITOR Charity Robey
PRODUCTION EDITOR Ken Santor
DESIGNER Harry Nolan
MANUFACTURING MANAGER Dorothy Sinclair
COVER ILLUSTRATION Harvey Chan

This book was set in 10/12 Times Roman by Publication Services, Inc., and printed and bound by Port City Press. The cover was printed by Lehigh Press, Inc.

Recognizing the importance of preserving what has been written, it is a policy of John Wiley & Sons, Inc. to have books of enduring value published in the United States printed on acid-free paper, and we exert our best efforts to that end.

The paper on this book was manufactured by a mill whose forest management programs include sustained yield harvesting of its timberlands. Sustained yield harvesting principles ensure that the number of trees cut each year does not exceed the amount of new growth.

Library of Congress Cataloging in Publication Data:
Beer, David F.
A guide to writing as an engineer / by David Beer and David McMurrey

p. cm
Includes index.
1. Technical writing. I. McMurrey, David A. II. Title
T11.B396 1997

808'.0666--dc20 96-13162
ISBN 0-471-11715-3 (pbk. : alk.paper) CIP

Printed in the United States of America

10 9 8

PREFACE

A Guide to Writing as an Engineer is intended for professional engineers, engineering students, and students in other technical disciplines. It not only addresses important writing concepts that apply to professional engineering communication, but also deals with the content, organization, format, and style of specific kinds of engineering writing such as reports, business letters, office memoranda, and e-mail. The book also covers oral presentations and how to find engineering information, both in the traditional ways and on the Internet.

WHO THIS BOOK IS FOR

The idea for this book grew from our experience in industry and the engineering classroom, and also from our wish to write a text that is practical rather than theoretical, and that devotes all its pages to the writing concerns of working engineers and those planning to become engineers. A common complaint among engineers and engineering students is that there is no helpful book on writing aimed specifically for them. Most technical writing texts focus, as their titles imply, on the entire field of technical writing. In other words, they aim to prepare readers and students for a complete knowledge of everything a technical writer is called on to do: they train people to become technical writers.

Engineers need to know how to write as much as anyone, but few have time to become technical writers. They are required to write all kinds of short documents and help in writing a variety of longer ones, but few need to acquire the skills of an advanced copy editor, graphic artist, or publisher. For most, engineering is their focus, and although advancement to management might bring increasing communication-related responsibilities and opportunities, these will for the most part still be focused on engineering and closely related disciplines.

Thus, we have written this book so that engineers and engineering students will have a resource that stays close to the real concerns they have in their everyday professional life. These are concerns we have identified over our combined thirty years of teaching and working in industry. Our perspective is the reason we give short shrift to some topics a technical writing book might spend several pages on, yet devote a chapter or two to what a traditional text might relegate to an appendix. These choices and priorities reflect what we have found to be important to the audience of this book—engineers and students of technical disciplines.

Our book is also written with the classroom in mind. It can serve as a text in a writing course for science and engineering majors, or indeed for any student who wants to become familiar with writing in the technology professions. Teachers will find the exercises at the end of each chapter good starting points for discussion and homework. Others who use the book will find these exercises well worth thinking about since they are designed to open up the material in the chapters to a larger context than the individual's own experience. The chapters themselves can be read from beginning to end, of course, but readers can also rely on the Table of Contents and Index to get them where they need to go. Thus the book can function not only as a textbook but also as a reference for engineering writing and research, and in the case of one chapter, for giving oral presentations.

WHAT IS IN THIS BOOK

To keep our book focused on the concerns of engineers, engineering students, and students of technical disciplines, we have organized the chapters in the following way:

Chapter 1, "Engineers and Writing," describes the importance of writing in your professional engineering life and provides a conceptual framework for understanding what impedes the communication process.

Chapter 2, "Some Guidelines for Good Engineering Writing," reviews a dozen essential requirements and guidelines for producing effective engineering documents.

Chapter 3, "Eliminating Sporadic Noise in Writing," reviews specific writing problems that can cause communication problems in engineering writing.

Chapter 4, "Writing Letters, Memoranda, and Electronic Mail," moves from the conceptual foundations covered in the preceding chapters to one of the most important applications of writing: professional correspondence. This chapter covers format and style for office memoranda, business letters, and e-mail. The chapter has a special section devoted to professional communications on the Internet.

Chapter 5, "Writing Some Common Engineering Documents," provides content, format, and style recommendations for such common engineering documents as inspection and trip reports, laboratory reports, specifications, progress reports, proposals, instructions, and recommendation reports.

Chapter 6, "Writing an Engineering Report," provides a standard format for the engineering report, with special emphasis on content and style for components such as the cover, transmittal letter, title page, table of contents, executive summary, graphics, tables, and documentation.

Chapter 7, "Accessing Engineering Information," outlines strategies you can use to find information in traditional libraries as well as on their contemporary online counterparts. This chapter contains a special section on finding information and using resources available on the Internet.

Chapter 8, "Engineering Your Presentations," reviews strategies you can use to prepare and deliver technical presentations, either individually or as part of a team.

Chapter 9, "Writing to Get an Engineering Job," covers the content, organization, style, and fornat for application letters and resumes—some of the main tools you'll use for getting engineering jobs.

Chapter 10, "Writing with Computers," provides an overview of features to look for in word-processing software, presents some tips for writing effectively with computers, and discusses other software you should know about if you use computers for much of your professional work.

ACKNOWLEDGMENTS

Many talented people have played a part, directly or indirectly, in bringing this book to print. We appreciate the input of recent students in the Department of Electrical and Computer Engineering at the University of Texas at Austin who are now successfully in industry or graduate school, including Steven Anderson,

Steve Burns, Wayne Contello, Luis Garcia, Adil Husain, David Kirk, Blane Leuscher, Joshua Massera, Valentin Medina, Jr., Jennifer Nossaman-Gropper, Albert A. Saenz, Cynthia Ann Torres, and Jose A. Vargas.

We also are grateful for the help of Susan Ardis, Head Librarian, Engineering Library, University of Texas at Austin; Professor Karin Ekberg, Technology Department, MidSweden University, Sweden; Professor W. Mack Grady, Department of Electrical and Computer Engineering, University of Texas at Austin; Leo Little, Telecommunications Engineer, 3M, Austin; Virgil Massey, C.P.A., professional business consultant; David Miller, software engineer, formerly of IBM Corporation; Jim Reidy, graphic design specialist at IBM Corporation; and Randy Schrecengost, an Austin-based professional engineer.

And of course we thank our families for their love, understanding, and support while we were writing this book: Ruth, Natasha, Phoebe, Patrick, and Jane.

David Beer
David McMurrey

Austin, Texas 1996

1

CONTENTS

1. Engineers and Writing 1

2. Some Guidelines for Good Engineering Writing 11

1

ENGINEERS AND WRITING

Communication skills are extremely important. Unfortunately, both written and oral skills are often ignored in engineering schools, so today we have many engineers with excellent ideas and a strong case to make, but they don't know how to make that case. If you can't make the case, no matter how good the science and technology may be, you're not going to see your ideas reach fruition.

George Heilmeier, corporate executive of Bellcore, in "Educating Tomorrow's Engineers," *ASEE Prism,* May/June 1995, p. 12.

Many engineers and engineering students dislike writing. After all, don't we go into engineering because we want to work with machines, instruments, and numbers rather than words? Didn't we leave writing behind us when we finished freshman English? The blunt fact, however, is that to be a successful engineer you must be able to write (and speak) effectively. If you could set up your own lab in a vacuum you might be able to minimize your first-hand communication with others, but all your ideas and discoveries would remain useless if they never got beyond your own mind.

If you feel you haven't mastered writing skills in college, the fault probably is not entirely yours. Few engineering colleges offer adequate (if any) courses in technical writing, and many students find what writing skills they did possess are badly rusted from lack of use by the time they graduate with an engineering degree. Ironically, most engineering programs devote less than 5% of their curriculum to communication skills—the very skills that many engineers will use some 20 to 40% of their working time. Even this percentage usually increases with promotion, which is why many young engineers spend the first five years of their careers wishing they had taken more math in college, and the second five wishing they had taken more writing courses.

But rather than dwell on the negative, let's look at the needs and opportunities that exist in engineering writing, and then see how you can best remove barriers to becoming an efficient and effective writer. You'll soon find that the skills you need to write well are no harder to acquire than many of the technical skills you have already mastered as an engineer or engineering student. First, here are four factors to consider.

ENGINEERS WRITE A LOT

Many engineers spend over 40% of their work time writing and usually find the percentage increases as they move up the corporate ladder. It doesn't matter that much of this writing is now sent through electronic mail (e-mail); the need for clear and efficient prose is the same whether it appears on a computer screen or sheet of paper. Much written material first read on a screen ends up being printed out on paper anyway—and the possibility of a paperless office, workshop, or engine room still seems pretty remote.

An engineer told us some years ago that while working on the B-1b bomber, he and his colleagues calculated that all the proposals, regulations, manuals, procedures, and memos the project generated weighed almost as much as the bomber itself. Most large ships carry several tons of maintenance and operations manuals. Two trucks were needed to carry the proposals from Texas to Washington for the ill-fated supercollider project. John Naisbitt estimated in *Megatrends* back in 1982 that some 6000 to 7000 scientific articles were being written every day, and even then the amount of recorded scientific and technical information in the world was doubling every five and a half years.

Who generates—that is, writes—all this material, together with countless memos, letters, reports, and numerous other documents? Engineers. Perhaps they get some help from a technical editor if their company employs one, and secretaries may play a large part in some cases. Nevertheless, the vast body of technical information available in the world today has its genesis in the work and reporting of engineers, whether they write as individuals or collaboratively in teams. Figure 1-1 shows the response we got when we asked just one electrical engineer friend at random to outline one of his days at 3M, a large, well-known company (the italics indicate where communication skills are called for).

7:30	Arrive, *read and respond to 3 Internet messages*
8:15	*Meet with* marketing on my ITU *paper abstract* to be given in Geneva
9:00	*Meet* with another engineer to *write specifications* for our new product
10:00	*Open and respond to letters and faxes*
10:30	*Give presentation* to Management Committee on NTT project status
11:00	Complete *expense report* on last trip
11:30	Lunch
12:00	Work on computer programming for PC interface project
2:00	*Read and respond to 6 or 7 intra-company E-mails*
2:30	Continue PC interface project
4:00	Prepare for Hydrological Testing *meeting*
5:00	Leave for the day

Figure 1-1 The working day of a typical engineer calls for plenty of communication skills.

ENGINEERS WRITE MANY KINDS OF DOCUMENTS

Few engineers work in a vacuum. Throughout your career you will interact with a variety of other engineering and nonengineering colleagues, officials, and members of the public. Even if you don't do the actual engineering work, you may have to explain how something was done, should be done, has to be changed, needs to be investigated, and so on. The list of all the possible engineering situations and contexts in which communication skills are needed is unending. Figure 1-2 on the next page identifies just some of the documents you might be involved in producing during your engineering career.

As we've already indicated, the delivery of written communication through e-mail is rapidly increasing as the twentieth century draws to its end. Used for anything from quick, pithy notes and memos to complete multipage documents, e-mail may before long become the most popular form of written communication. Although this won't change the need for clarity and organization in engineering writing, some conventions may develop that we will need to adjust to. Whatever the future holds, a solid skill in clear and efficient writing, and the ability to adapt to many different document specifications, will probably be necessary for as long as humans communicate with each other.

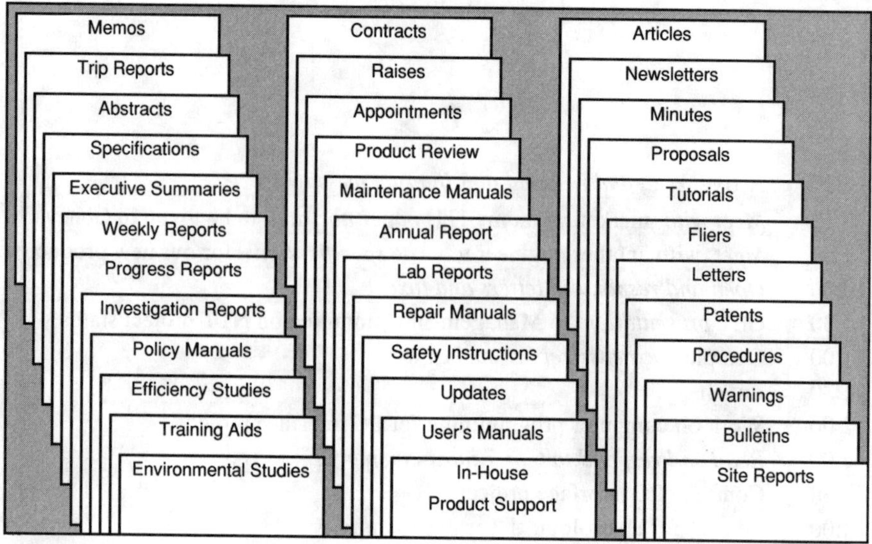

Figure 1-2 Engineers work on many kinds of written documents throughout their careers.

A SUCCESSFUL ENGINEERING CAREER REQUIRES STRONG WRITING SKILLS

In the engineering field you are rarely judged solely by the quality of your technical knowledge or work. People also form opinions of you by what you say and write. When you write a memo or report, talk to members of a group, deal with vendors on the phone, or attend meetings, the image others get of you is largely formed by how you communicate. Even if you work for a large company and don't see a lot of high-level managers, they can still gain an impression of you by the quality of your written reports as well as by what your immediate supervisor tells them.

Writing on "What Courses Should I Take?" in a newsletter for engineering students at the University of Texas at Austin, Richard C. Levine, Manager of Hardware Planning at Bell Northern Research stated:

> *Another fundamental is the ability to read with comprehension and to write clearly and correctly. . . . I can't emphasize enough that both of these skills are extremely important. I am not a picky person when it comes to spelling and grammar, but when I see a report or memo that has repeated errors I*

immediately question the ability and dedication of the person who wrote it. Why didn't they take the time and effort to do it right? Most of the successful engineers I know write clear, well-organized memos and reports. Engineers who can't write well are definitely held back from career advancement.

The Current, vol. 3, no. 1, (April 1987), p. 1.

Opinions like this are common among engineering management. The majority of people who have advanced in an engineering career will tell you the same thing. If you don't believe us, ask them!

Two relatively recent trends, **accountability** and **specialization**, are now making communication skills even more crucial in the engineering profession. Probably more than ever before, engineers and their companies are being held accountable by the public. People want to know *why* a space shuttle crashed (after all, their taxes paid for the mission). They want to know if it really is safe to live near a nuclear reactor or high-power lines. The public—often through the press—wants to know if a plant is environmentally sound or if a project is likely to be worth tax dollars. Moreover, there is no shortage of lawyers ready to hold engineering firms and projects accountable for their actions.

All of this means that engineers are being called upon to explain themselves in numerous ways, to an increasing variety of people—many of whom are not engineers. Here are the words of Norman Augustine, chairman and CEO of Martin Marietta Corporation and also chair of the National Academy of Engineering:

Living in a "sound bite" world, engineers must learn to communicate effectively. In my judgment, this remains the greatest shortcoming of most engineers today—particularly insofar as written communication is concerned. It is not sensible to continue to place our candle under a bushel as we too often have in the past. If we put our trust solely in the primacy of logic and technical skills, we will lose the contest for the public's attention—and in the end, both the public and the engineer will be the loser.

Norman R. Augustine, in *The Bridge,* The National Academy of Engineering, vol. 24, no. 3, Fall 1994, p. 13.

While the public demands that engineering concepts be made more understandable to laypeople, engineers are also finding it increasingly difficult to communicate with one another. Almost daily, engineering fields once considered unified become progressively fragmented. Thus it's quite possible for two engineers with similar degrees to have large knowledge gaps when it comes to each other's work. This means in practical terms that a fellow engineer may have little more understanding of what you are working on than does a layperson. These gaps in knowledge often have to be bridged, but can't be unless specialists have the skills to communicate clearly and effectively.

ENGINEERS CAN LEARN TO WRITE WELL

Writing is not easy for most of us, and like programming, painting, or playing the piccolo, good writing takes practice. A lot of truth lies in the adage that no one can be a good writer—only a good *re*writer. If you look at the early drafts of the most famous authors' works, you will see various changes, additions, deletions, rewordings, and corrections. So don't expect to produce a perfect piece of writing on your first try. Every first draft of a document, whether it's a one-page memo or a fifty-page set of procedures, needs to be worked on and improved before being sent to its readers.

As an engineer you have been trained to think logically. In the laboratory or workshop you are concerned with precision and accuracy. From elementary and secondary school you already possess the skills needed for basic written communication, and every day you can see samples of clear writing in newspapers, weekly news magazines, and articles in popular journals. Thus you are already in a good position to become an effective writer simply by recognizing and structuring what you've already been exposed to. All you need is some instruction and practice. This book will give you plenty of the former, and your engineering career will give you many opportunities for the latter.

NOISE AND THE COMMUNICATION PROCESS

Have you ever been annoyed by interference on your television screen during a favorite program? Perhaps a neighbor was talking on CB radio, and the transmission did nasty things to your reception. Or maybe you couldn't hear a friend clearly on the telephone because someone was using the vacuum cleaner in the next room or the stereo was booming.

In each case, what you were experiencing was noise interfering with the transmission of information. Whenever a message is sent, someone is sending it

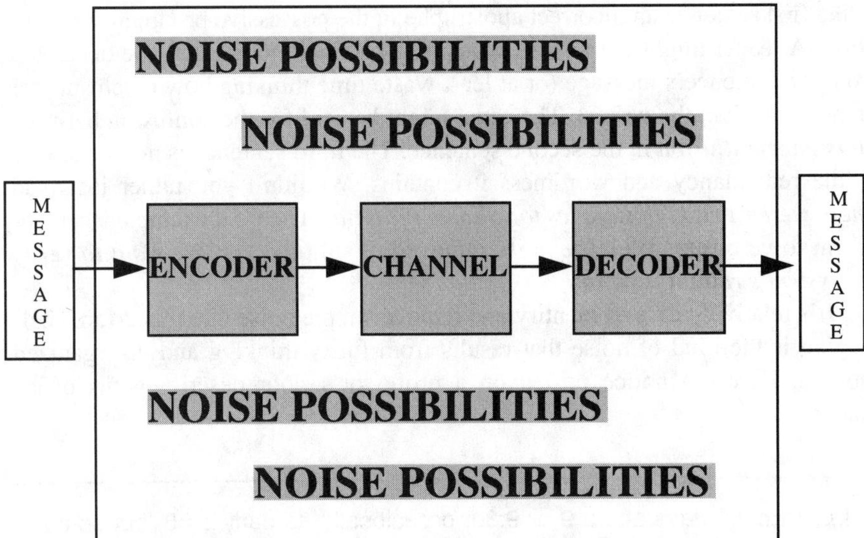

Figure 1-3 In noise-free technical communication, the message moves from the encoder (writer, speaker) to the decoder (reader, listener) without interruption or ambiguity.

and someone else is trying to receive it. In communication theory, the sender is the *encoder,* and the receiver is the *decoder*. The message is sent through a channel, usually speech, writing, or some other conventional set of signals, and anything that prevents the message from flowing clearly through the channel from the encoder to the decoder is *noise*. Figure 1-3 illustrates this concept. Note how all our actions involving communication are "overshadowed" by the possibility of noise.

Applying this concept to engineering writing, we can say that anything causing a reader to hesitate in uncertainty, confusion, or frustration, is noise. The following sentences contain just a few samples of written noise:

> When they bought the machine they weren't aware of it's shortcomings.
>
> They were under the allusion that the project could be completed in 6 weeks.
>
> There was not a sufficient enough number of samples to validate the data.
>
> Our intention is to implement the verification of the reliability of the system in the near future.

In the first sentence an incorrect apostrophe in the possessive pronoun *its* causes noise. A reader might "trip over" this apostrophe and momentarily be distracted from the sentence's message (or at least waste time thinking how much smarter he or she is than the writer). The same might be said for the confusion between *allusion* and *illusion* in the second sentence. The third sentence is noisy because of the redundancy and wordiness it contains. Wouldn't you rather just read *There weren't enough samples to validate the data?* The final example is a monument to verbosity. With the noise removed, it simply says: *We want to verify the system's reliability soon.*

It's relatively easy to identify and remove simple noise like this. More challenging is the kind of noise that results from fuzzy thinking and disorganized thought. Here's a notice posted on a professor's door describing his office hours:

> I open most days about 9 or 9:30, occasionally as early as 8, but some days as late as 10 or 10:30. I close about 4 or 4:30, occasionally around 3:30, but sometimes as late as 6 or 6:30. Sometimes in the mornings or afternoons I'm not here at all, but lately I've been here just about all the time except when I'm somewhere else, but I should be here then, too.

Academic humor, maybe, but it's not hard to find writing in the engineering world that is equally difficult to interpret, as this excerpt from industrial procedures shows:

> *If containment is not increasing or it is increasing but MG Press is not trending down and PZR level is not decreasing, the Loss of Offsite power procedure shall be implemented, starting with step 15, unless NAN-S01 and NAN-S02 are de-energized in which case the Reactor Trip procedure shall be performed. But if the containment THRSP is increasing the Excess Steam Demand procedure shall be implemented when MG Press is trending down and the LIOC procedure shall be implemented when the PZR level is decreasing.*

The following chapters contain advice, illustrations, and strategies to help you avoid such writing. Meanwhile, try to keep the concept of noise in mind when you write or edit, whether you are working on a five-sentence memo or a 500-page technical manual. Throughout your school years you may have been reprimanded for "poor writing," "mistakes," "errors to be corrected," and so on, but as an engineer it might be better to think in terms of *noise to be eliminated from written communication.* For efficient and effective communication to take place, the signal-to-noise ratio must be as high as possible.

CONTROLLING THE WRITING SYSTEM

Engineers frequently design, build, and manage systems made up of interconnected parts. Controls have to be built into such systems to guarantee that they function correctly and reliably and produce the desired result. The machinery used to mill propeller shafts for large ships must be guided by a control system to ensure that correct tolerances and other specifications are met. If the ATM machine chews up your credit card and spits it back out to you in place of the $50.00 you had hoped for, you'd claim the system is not working right, or that it is out of control. The system is only functioning reliably if the input (your credit card) produces the desired output (your $50.00).

What has this got to do with writing? Well, we can view language as a *system* made up of various components such as sounds, words, clauses, sentences, and so on. Whenever we speak or write, we use this system, and like other systems it must be controlled if it is to do its job right. The person who supposedly wrote in an accident report, *Coming home, I drove into the wrong house and collided with a tree I didn't have,* was obviously unable to express what really happened. The input (thought) to the system (language) did not have the desired output (meaning) because the writer was not in control of the system.

In the same way, an instruction like *Pour the concrete when it is above 40°F* indicates a lack of language control since the writer is not clearly stating whether the concrete or the weather must meet the specification of "above 40°F." Thus you might think of language as a system or even a tool you can learn to control so that it will do exactly what you want it to. Learning to control language, namely to write and speak so you get desired results or feedback, is really not much different than training yourself to operate complex machinery or software systems. With some help and effort you can train yourself to eliminate most if not all noise that might occur when you transfer information by means of writing and speaking.

If you get the response you want from your communication, you can be pretty sure you have communicated well. A proposal accepted, a part promptly delivered, a repair quickly made, an applied-for promotion awarded—these are just a few examples of the payback from effective communication. To put it another way, if you learn to efficiently control the tool you are using (language) so that it's noise free, you will produce clear and effective written documents that get results.

EXERCISES

1. Interview three professional engineers concerning the amount and kinds of writing they do on the job. How much of their time is spent writing each day? Is the amount of writing they do related to how long they have been with their company? In what ways do they feel that their writing skills have helped (or hindered) them in their careers so far? Do they get any help with their writing from secretaries, peers, or technical writers? What is the attitude of their superiors toward clear writing?
2. Look at the list of technical documents in Figure 1-2. How many are you familiar with? Can you think of examples of some of these documents? When would they be likely to be important to you as a reader? Can you think of other types of documents not included in Figure 1-2? Ask some engineering friends how many kinds of documents they have worked on, either as individuals or as part of a group.
3. Think of your own engineering major or specialty. List some engineering fields most closely related to yours, some that are marginally related, and some that are only remotely related. What kinds of technical knowledge do you share with people in these fields? At what point is your common knowledge likely to be no longer useful? What problems can you foresee in communicating technical information with engineers in other fields? What problems would you face if you had to talk about your field to a nonengineering audience?
4. As we point out in this chapter, noise is anything that interferes with efficient transmission of information. We've all experienced noise when trying to communicate with another person—and most of us have at times created it. What kinds of noise do you think you create in your written communication? Is it primarily in your spelling, grammar, sentence structure, organization of thoughts, or what? How about in your spoken communication? What kinds of noise sometimes interfere with your receiving and understanding the written or spoken communication of others?

BIBLIOGRAPHY

Anderson, Paul V. "What Survey Research Tells Us about Writing at Work," in *Writing in Nonacademic Settings,* eds., Lee Odell and Dixie Goswami. New York: Guilford Press, 1985.

Angell, David, and Brent Heslop. *The Elements of E-Mail Style.* Reading, MA: Addison-Wesley, 1994.

Arthur, Richard H. *The Engineer's Guide to Better Communication.* Glenview, IL: Scott, Foresman and Company, 1984.

Nagle, Joan G. *Handbook for Preparing Engineering Documents.* Piscataway, NJ: IEEE Press, 1996.

Naisbitt, John. *Megatrends.* New York: Warner Books, Inc., 1982.

2

SOME GUIDELINES FOR GOOD ENGINEERING WRITING

This chapter presents guidelines for writing efficiently and producing useful documents. Although different people approach writing tasks in somewhat different ways, these guidelines follow in general the overall process used by successful engineering writers. We have also focused on these topics because they represent common problems you as an engineer are likely to have in the process of writing and formatting your documents. This chapter assumes, of course, that you already have the material or topic that you need to write about.

FOCUS ON WHY YOU ARE WRITING

Before starting to write you should have a good idea of precisely what you want to communicate to your audience. If these goals aren't first defined in your own mind, you can't really expect your readers to get a clear message. Having this sense of purpose as you write may not guarantee your readers will receive a noise-free message, but writing without a clear goal will almost certainly result in poor communication. Thus, whether you have to write a short memo or a lengthy technical report, you should start with a firm sense of purpose so you can present supporting data, test its adequacy, and discard anything that is not needed.

Broadly speaking, the purpose of most technical writing is either to present information or to persuade people to act or think in a certain way. Frequently your documents will have to be both informative and persuasive. To fine-tune your sense of purpose before writing, ask yourself the following:

Do I want to

1. **inform**—to provide information without necessarily expecting any action on the part of my reader(s)?
2. **request**—to obtain permission, information, approval, help, or funding?
3. **instruct**—to give information in the form of directions, instructions, procedures, or the like, so my readers will be able to do something?
4. **propose**—to suggest a plan of action or respond to a request for a proposal?
5. **recommend**—to suggest an action or series of actions based on alternative possibilities that I've evaluated?
6. **persuade**—to convince or "sell" my readers, or to change their behavior or attitudes based on what I feel to be valid opinion or evidence?
7. **record**—to document for the record how something was researched, carried out, tested, altered, or repaired?

How you write any document should be guided by what you want your audience to do with your information, and what they need from the document in order to be able to do it. Thus, your audience plays a definitive role in determining how you approach your task. Do they need to be informed, instructed, dissuaded, or what? Only a careful analysis of your purpose or purposes for writing and the nature of your audience can give you the answers and thus enable you to write to the point.

FOCUS ON YOUR READERS

If you found yourself in a remote region and met people who had never seen anything electronic, you wouldn't hand them your scientific calculator and expect them to use it. First a great deal of technology transfer would have to take place; you would have to teach your "audience" how to use the calculator (assuming they cared to know). This may seem obvious, but a lot of technical writing fails because writers make inaccurate assumptions regarding the people who read their documents. We often write without taking adequate time initially to consider the needs, interests, levels of expertise, or possible reactions of those who must read our work.

This is not just a question of being polite, thoughtful, or sensitive. Since your goal is to send a clear, noise-free message through your document to your audience, you must consider their abilities and expectations as you plan, write, and revise. As an engineer, you may find yourself writing to a variety of people either in your immediate group, close by in the company, elsewhere in the company, or outside the company. Sometimes you will write to your professional and technical peers, sometimes to your superiors, and other times to those "below" you. In all these writing situations, inadequate audience analysis will inevitably result in noise.

No matter who you write to, you write because you expect some kind of resulting action, even if it is only nonphysical "action" such as permission, understanding, or a change of opinion. To get results, your communication must bridge a gap between you and your target audience. In the working world this gap is likely to be caused by variations in **knowledge**, **ability**, or **interest**. Obviously, the three may overlap, but to determine where you stand before putting any effort into writing, first identify who your audience is and then ask yourself these questions:

Knowledge

- Are my readers engineers in my field of expertise who are seeking technical information, and will they be offended or bored by elementary details?
- Are they engineers from a different field who will need some general technical background first?
- Are they managers or supervisors who may be less knowledgeable in my field but who need to make executive decisions based on what I write?
- Are they technicians or others without my expertise and training but with a strong practical knowledge of the field?
- Are they nonexperts from marketing, sales, finance, or other fields who lack engineering or technical backgrounds but who are interested in the subject for nonengineering reasons?
- Are they a mixed audience, such as a panel or committee, made up of experts and laypeople?

Ability

- Am I communicating technical information on a level my audience can use?
- Am I using appropriate vocabulary, examples, definitions, and depth of detail?
- Am I expecting more expertise, skill, or action from my audience than I can reasonably expect?

(continued)

Interest

- Why will my audience want to spend time reading this document?
- Does my document provide the right level of detail and technology to keep my audience's interest without losing them or boring them?
- What is their current attitude likely to be—positive, neutral, or negative?
- Will my document give them the information they want?

The answers to these questions will increase your awareness of the multiple decisions and choices to be made as you plan, write, or revise your document. Remember, in order to deliver a clear message, you should first assess your audience. You need to know who you are writing to and have a clear idea of their technical knowledge, expectations, and attitude toward the subject. If you properly analyze these and address them in your document, you are well on your way to communicating effectively.

SATISFY DOCUMENT SPECIFICATIONS

Before writing, you should be aware of any specifications your document must meet. Many audiences expect documents they receive to be within certain parameters. If management asks for a brief memo, they may be irritated when you overload their circuits with a lengthy, detailed treatise. When a technician requests the specs on a frequency tester, it won't be appreciated if you come up with a flowery prose discussion on the strengths and weaknesses of the equipment. If you respond to an RFP (request for proposal) that calls for a proposal of no more than ten pages but submit something twice that long, chances are your proposal will be eliminated from the competition.

Various document specifications exist. Such specifications may require you to provide sections addressing certain topics in your report, such as experimental problems, environmental impact, decisions reached, budget, and so on. The editors of an engineering journal may put limits not only on the number of words but also the number of graphics your technical paper can include. A word limit is frequently placed on the length of an abstract or summary. Many reports have requirements not only for their length but also for such matters as headings, spacing, and margin width. Some government agencies require that the proposals they receive be written in specific formats, in certain fonts, and even with restrictions on how many letters are permitted in each line of text. Here is an example from an RFP for a government research program:

> Each proposal shall consist of not more than five single spaced pages plus a cover page, a budget page, a summary page of no more than 300 words, and a page detailing current research funding. All text shall be printed in single-column format on 8-1/2 × 11-inch paper with margins of at least 1 inch on all sides. . . .

Knowing precisely what is expected of you *before* you begin to write will prevent wasted time and give your document a better chance at success.

GET TO THE POINT

Anyone reading your memos, letters, and reports is likely to be in a hurry. Few engineers have the leisure for "biblical" reading—where one reads from Genesis to Revelation to discover how things turn out. Just as your sentences need to be direct, your documents need to have the most important information at the beginning. This means moving from the general to the specific. Readers would much rather know your key points, complaints, requests, conclusions, or recommendations before they read supporting details. For instance, if you do a series of tests to determine whether some equipment should be replaced, your supervisor will want to know what you have found out and what you recommend. A complete, detailed description of your test procedures may be necessary to support your main points and will likely be verified by others—but it could go unread by those in management who need only the "bottom line."

Where you tell your readers what they most need to know depends on the kind of document. In a letter it will be in the opening sentences. In a memo you should provide a subject line making more than just a vague reference to the overall topic. Look at these examples:

> Vague: *SUBJECT: Employee safety*
> Better: *SUBJECT: Need for employees to wear hard hats and safety glasses*
>
> Vague: *SUBJECT: Emergency requisitions*
> Better: *SUBJECT: Recommendations to change the procedures for making emergency requisitions*

A lot of memos are now sent by e-mail, which may limit the number of characters for your subject heading. If this is the case, then the challenge is to get as much meaning as possible into a small space and to clearly state your key message in the opening sentences of the memo.

In a longer report your main points should become quickly evident to your reader through an informative title followed by a summary of your findings, conclusions, recommendations, results, or whatever the important information is. (See the chapters on individual reports and the sections on abstracts and executive summaries in this book.) No matter what kind of document you are producing, however, first determine your audience and purpose, and then give your readers the information they most need in the place they can most efficiently access it—the beginning of the paper, rather than buried somewhere in the middle or at the end.

PROVIDE ACCURATE INFORMATION

Even the clearest writing is useless when the information it conveys is wrong. If you state that an ampere can be defined as a coulomb of charges passing a given point in 10 seconds rather than 1 second, you have presented wrong information. If you refer to data in Appendix B of your report when you mean Appendix D, the error could stump your readers and cause them to lose confidence in your report. Inaccurate references to the work of others will cause your readers to be highly suspicious of the reliability of your entire report—and even of your honesty as a writer. Inaccurate directions in a set of instructions or procedures might be frustrating at best, disastrous at worst. Another kind of inaccuracy might be a claim that is true sometimes but not under all conditions, for example that water always boils at 100°C. What about purity and variations in atmospheric pressure?

There is also a great difference between **fact** and **opinion**. A fact is a dependable statement about external reality that can be verified by others. An opinion expresses a feeling or impression that may not be readily verifiable by others. The danger comes when opinions are stated as facts. Note the difference between these two:

Fact: *The NR-48 tool features multiple programmable transmitters and a five-station receiver array.*

Opinion: *The NR-48 is by far the best piece of equipment for our purposes.*

The second statement might be "correct" but is still only an opinion until supported by verifiable facts. To be strictly honest the writer should identify it as an opinion unless evidence is presented to support it as fact. In short, make sure that (1) your facts are correct when you write them down and (2) your opinions are presented as such until adequate evidence is provided to verify them.

PRESENT YOUR MATERIAL LOGICALLY

Not only should it be easy to access your document's essential message, but all your information should be in the right place. This means you should organize your material so that each idea, point, and section is clearly and logically laid out within an appropriate overall pattern. If you are following document specifications provided by someone else, you have little choice but to follow those specs, but even within a prescribed plan of organization you may have some leeway to present material the way you feel is most effective.

As always, think before writing, and keep your readers firmly in mind. If they want to know what progress you have made on a project, what you did on a trip, or how to carry out a procedure, obviously they will expect your material to be in chronological order. If they are expecting a description of a piece of equipment or of the layout of some facilities, they should be provided with a description that logically moves from one physical point to another.

On the other hand, if you have a number of points to make such as five ways to reduce costs or six reasons why a project must be canceled, present those points from the most to the least important, or vice versa. Perhaps your material needs to be presented in order of familiarity or difficulty, as when you are writing a tutorial or textbook. Or you may want to move from the general to the specific, as when you write a memo first stating that more stringent safety regulations are needed at your plant and then provide examples of current unsafe practices.

EXPRESS YOURSELF CLEARLY

Engineering is considered a precise discipline—although in reality, as most engineers will admit, it's not always as precise as we would like it to be. Machine parts, for example, may be allowed a certain degree of variation or tolerance within a specified zone and still be interchangeable. Similarly, you have some choices in how you express yourself in engineering writing. You can often say the same thing three or four different ways, but your overriding concern should always be to state what you have to say clearly and to the point. Don't force your readers to work harder than necessary to grasp what you have written; your sentences must convey a single meaning with no room for interpretation or misunderstanding. If your readers yearn for uncertainty and suspense, they can read a romantic novel or detective story; if they enjoy different connotations and levels of meaning, they can read poetry. So, here are some pitfalls to avoid.

AMBIGUITY

The word *ambiguous* comes from a Latin word meaning to be undecided. Ambiguity primarily results from permitting words like *they* and *it* to point to more than one possible referent in a sentence, or from using short descriptive phrases that could refer to two or more parts of a sentence. In either case, your reader becomes confused—undecided—and may interpret your sentence differently than you intended, as illustrated below.

Ambiguous: *Before accepting materials from the new subcontractors, we should make sure they meet our requirements.*

(Who are *they,* the materials or subcontractors?)

Clear: *We should make sure the materials from the new subcontractors meet our requirements before we accept them.*

Ambiguous: *The microprocessor interfaced directly with the 7055 RAM chip. It runs at 5 MHz.*

(What does *it* refer to?)

Clear: *The microprocessor interfaced directly with the 7055 RAM chip. The 7055 runs at 5 MHz.*

Ambiguous: *Our records now include all development reports for B-44 engines received from JPL.*

(What was received from JPL—the reports or the engines?)

Clear: *Our records now include all B-44 engine development reports received from JPL.*

Ambiguous: *After testing out at the specified high temperatures, the company accepted the new chip.*

(Did the company or the chip test out at the high temperatures?)

Clear: *The company accepted the new chip after it tested out at the specified high temperatures.*

VAGUENESS

If ambiguity causes readers to see more than one meaning in your writing, vagueness causes them to see no useful meaning at all. What would you think if your doctor told you to "take a few of these pills every so often"? You would want him or her to provide some facts and figures. Explanations or directions lacking specific detail sound fuzzy and unfocused, more like personal opinion than useful data.

Abstract words are not inherently wrong, but they fail to provide the precision effective technical writing needs. Try to avoid abstract words and phrases like *pretty soon, substantial amount,* and *corrective action,* and replace them with terms that have exact meaning such as *in three days, $8,436.00, replace the altimeter.* Here are two more examples of vague writing and ways they can be remedied:

Vague:	*The Robotics group is several weeks behind schedule.*
Useful:	*The Robotics group is six weeks behind schedule.*
Vague:	*The CF553 runs faster than the RG562 but is much more expensive.*
Useful:	*The CF553 runs 84% faster than the RG562 but costs $4,320.00—$2,840.00 more than the CF553.*

As you can see in the second example, vague writing might require fewer words, but it's rarely wise to be concise at the expense of precision. This is especially true when writing instructions and specifications.

COHERENCE

The root of the word *coherence* is *cohere*, meaning to stick together, and as you know, a cohesive does just that. Coherence in writing means how well paragraphs and even complete documents "stick together"—that is, how well they stay focused on their true subject. In a coherent paragraph, all the sentences clearly belong where they are because they address only the topic of the paragraph and are logically connected to one another. You might say the sentences stick to the point and stay there. Coherence in a complete report also means how well the report is designed to take the reader through its paragraphs and sections by means of clear transitions such as headings and subheadings, and how all the sections focus on and support the subject of the report. (Our chapters on report writing will show you how to achieve coherence in longer documents.)

You can achieve coherence in your paragraphs by making sure each sentence clearly relates to the one before it and after it. This means opening with your main point or topic sentence, repeating key words where needed, and using transitional words (see Ch. 3) and pronouns to link sentences as they build up the paragraph. Note how the following paragraph lacks coherence and how it is improved by the devices in boldface in the revised version.

Poor coherence:

A significant disadvantage of the 125-H CRT is its high power consumption. The tube requires substantial power to produce the high voltages and currents that are necessary to drive and deflect the electron beam. The 125-H is inefficient—only about 10% to 20% of the power used by the tube is converted into visible light at the surface of the screen. The 125-H is poorly suited for portable display devices that run on batteries, where lower power consumption is necessary. We should consider other options before committing to purchase the 125-H.

Effective coherence:

A significant disadvantage of the 125-H CRT is its high power consumption. ***This*** *tube requires substantial power to produce the high voltages and currents that are necessary to drive and deflect the electron beam.* ***In addition,*** *the 125-H is inefficient—only about 10% to 20% of the power used by the tube is converted into visible light at the surface of the screen.* ***Thus,*** *the 125-H is poorly suited for portable display devices that run on batteries, where lower power consumption is necessary.* ***Because of this drawback,*** *we should consider other options before committing to purchase the 125-H.*

DIRECTNESS

Being as direct as possible in your writing lets your reader grasp your point quickly. Suspense might be thrilling, but a busy technical reader wants access to your information quickly and easily. The most important part of your message should come at the beginning of your sentences and paragraphs. Here are some examples of what this means at the sentence level:

Indirect: *After a long and difficult development cycle due to factory renovation, the infrared controller will be ready for production in the very near future.*

Direct: *The infrared controller will be ready for production March 4. Its development cycle was slowed by the factory renovation.*

Indirect: *Fred has been busily working on this project. This past week he also reworked the logic diagrams, rewired the controller arm, and redesigned all of the RIST circuitry.*

Direct: *Fred redesigned the RIST circuitry on Thursday. He also reworked the logic diagrams and rewired the controller arm last week.*

USE EFFICIENT WORDING

Opinions vary on how much it costs a company for an employee to produce one written page of information, but between $15.00 and $20.00 is a reasonable current estimate. When you think of all the people writing letters, memos, reports, manuals, proposals, and countless other documents for industry, you see how the costs mount up. Add to this the fact that most of us have little training in producing concise prose, and you can appreciate how sharpening your writing and editing skills can mean not only saving time, but money. Moreover, since we all tend to be wordy, carefully editing our work can often reduce or eliminate a lot of time-consuming work for our readers.

WORDINESS

Using an unnecessarily pompous word instead of a straightforward one can cause your readers to slow down. Choose the simplest and plainest word whenever you can. Your readers can be distracted or even confused by words that call attention to themselves without contributing to meaning. This pitfall becomes even more likely if some of your readers are not native speakers of English, as is often the case in engineering fields today. Write to communicate rather than to impress, or as the saying goes, "Never utilize *utilize* if you can use *use*." A few of the more ostentatious—oops, make that showy—words found in engineering writing are listed here, with some plain, equally efficient counterparts:

commence	*start*	fabricate	*make*	proceed	*go*
compel	*force*	finalize	*end*	procure	*get*
comprises	*is*	initiate	*begin*	rendezvous	*meet*
employ	*use*	optimal	*best*	terminate	*end*
endeavor	*try*	prioritize	*rank*	visitation	*visit*

Wordiness can also result from using far more words than you need to express an idea. Unkind editors sometimes refer to this as verbiage (by analogy to garbage?). Few of us appreciate hearing

I regret to say that at this point in time I basically do not have access to that specific information...

when a simple "I don't know" is enough. Similarly, your reader is unlikely to thank you for having to plow through

It is our considered recommendation that a new computer should be purchased...

when you could have simply said you recommend buying a new computer. You can eliminate a lot of wordiness in your writing by training yourself to edit carefully and to make every word count. Look at the following three pairs; you will see which sentences are more efficient and noise free.

It is essential that the lens be cleaned at frequent intervals on a regular basis as is delineated in Ops Procedure 132-c.
Clean the lens frequently and regularly (see Ops Procedure 132-c).

The location of the experimental robotics laboratory is in room 212A.
The experimental robotics lab is in 212A.

There are several EC countries that are now trying to upgrade the communication skills of their engineers.
Several EC countries are trying to upgrade the communication skills of their engineers.

You can also reduce wordiness by avoiding certain pretentious phrases which have unfortunately become common. A good style book will give numerous examples, but here are a few that crop up frequently in engineering writing:

Verbiage	Efficient
a large number of	*many*
at this point in time	*now*
come in contact with	*contact*
exhibits the ability to	*can*
in the event of	*if*
in some cases	*sometimes*
in the field of	*in*
in the majority of instances	*usually*
in the neighborhood of	*about*
in view of the fact that	*because*
in view of the foregoing	*therefore*
serves the function of being	*is*
subsequent to	*after*
the reason why is that	*because*
within the realm of possibility	*possible*

Check your writing for such unnecessary phrases and for unneeded words in general, as we do in the next sentence. You may ~~often~~ find ~~that there are a number of~~ words ~~contained in your writing~~ that can be ~~safely~~ eliminated without any ~~kind of~~ danger to your meaning ~~whatsoever~~. If you let your writing "cool off" for a while and come back to edit later, chances are you will discover more wordiness than if you try to edit immediately after writing.

REDUNDANCY

One category of verbiage is redundancy. This means using words that say the same thing, like *basic fundamentals,* or phrases that duplicate what has already been said, as in *They decided to reconstruct a hypothetical test situation that does not exist*. In fact if you master the art of redundancy you can make everything you write almost twice as long as need be. A few common redundant pairs are identified below, but the list is far from exhaustive.

Redundant	Efficient
alternative choices	*alternatives*
actual experience	*experience*
completely eliminate	*eliminate*
component part	*component (or part)*
connected together	*connected*
collaborate together	*collaborate*
diametrically oppose	*oppose*
exactly identical	*identical*
integral part	*part*
just exactly	*exactly*
permeate throughout	*permeate*
prove conclusively	*prove*
rectangular in shape	*rectangular*
12 noon	*noon*
very best	*best*

Again, we all can be wordy at times, so it's a good idea to edit your writing once, simply looking for redundancy and wasted words. Grammar-checking software can help, but you still need human editing to remove this kind of noise from your writing.

TURNING VERBS INTO NOUNS

Replacing a perfectly good verb (action) with a noun (the name of an action) is unfortunately common in much engineering writing. This is often the result of wanting to write in the passive rather than active voice. Look at these two pairs of sentences:

> *An analysis of the data will be made when all the results are in.*
> *We will analyze the data when all the results are in.*
>
> *An investigation of all possible sources of noise was undertaken.*
> *All possible noise sources were investigated.*

It's easy to see which sentences are shorter and more natural. If you take the verb that really matters in a sentence (such as *analyze* and *investigate* in the preceding examples) and make a noun of it, you are forced to add another, generally weaker verb to convey your meaning.

Thus you will write *made a selection of* instead of *selected,* or *procurement of services can be accomplished by,* instead of *services can be procured by.* Note that many such verbs when changed into nouns need to be followed by *of.* Grammar checkers use this as a cue to warn you of the problem, but again, there is no better tool than your own editing skills—or those of a competent and honest colleague—to free your writing of verbiage.

MAKE YOUR IDEAS ACCESSIBLE

Without even reading a word, we can look at the pages of a document and get a good idea of how efficiently the material is presented. This impression comes from the structure of the material—specifically, how well the material is laid out in accessible "chunks" for the reader. The two most important factors here are (1) the subdivision of material into sections and subsections with hierarchical headings and (2) paragraph length.

HIERARCHICAL HEADINGS

Even in short engineering documents, a system of headings is essential to keep your material clearly organized and to let readers know what is in each section of the document. Headings and subheadings are also signposts that help a reader get through a report without getting lost. Moreover, they reveal the hierarchical relationships of your material, enabling readers to understand the various levels of detail or importance in your work. Clear and informative headings also help readers find the parts of your report that interest them most.

Although practice differs among engineers and organizations, a common format for the first three levels of headings is as follows:

FIRST-LEVEL HEADING

Write first-level headings in capital letters and put them flush with the left margin of the page. Use boldface to make the heading stand out, and separate it from the written material above and below it by at least one space, as in this illustration.

Second-Level Heading

Also place second-level headings flush with the left margin with at least one space separating them from any text. Capitalize only the first letter of each main word, and make these headings boldface. If you don't have access to boldface type you can underline your headings, although underlining does clutter the text. In any case, don't use both boldface and underlining for headings.

Third-level headings. Indent third-level headings and place them on the same line as the text they precede. They are capitalized as a sentence would be and can be in boldface or italics.

NUMBERED HEADINGS

Sometimes you may be required to add a numbered, or decimal, system to your headings, and in fact many companies and suppliers require such numbering. A number system gives readers easier reference to parts of a very long report. Note that the different levels of heading can be successively indented, although many companies don't follow this practice.

FIRST LEVEL	**1. 0 QUALITY ASSURANCE PROVISIONS**
Second level	**1.1 Contractor's Responsibility**
Third level	**1.1.1 Component and material inspection**
Fourth level	**1.1.2.1 Laminated material certification**

When you use this system, make sure it doesn't get out of control. If your material is so complicated or detailed that you are getting down to levels such as

2.11.3.4.6.23, as some manuals do, then maybe it's time to inspect your document closely to see where you can break it up into smaller, more manageable sections or short chapters, each with its own verbal heading and independent hierarchies within it.

These structural elements of a document (and again, it doesn't matter whether it's a two-page memo or a 500-page manual) can be planned ahead of time. Writing skills aren't needed so much for this as planning and outlining skills—plus an awareness that the headings, divisions, and subdivisions in your document play a vital part in making your information clear and easily available to your readers. So spend some time thinking about how you are going to arrange and format your document; this will help you avoid noise at the structural level before you even begin to write. You might, of course, still be able to improve the structure and organization of your paper after completing the first draft. Modern word processors make this easy and even enjoyable.

PARAGRAPH LENGTH

No one, especially in technical fields, wants to read a solid page of wall-to-wall text of difficult material. A busy manager, for example, will want to absorb your information in as easily digestible pieces as possible.

Dense text on a page creates noise simply because it's so discouraging. When your readers are trying to follow demanding technical information they are already challenged, and presenting it to them in solid page-long chunks is at least going to give them mental indigestion. Later, if they want to quickly find a point you made or a piece of data you presented, they are going to have trouble locating it if they have to wade through a ponderous paragraph.

As an example of what we mean, we have reformatted most of this section on paragraph length with no paragraph breaks:

> No one, especially in technical fields, wants to read a solid page of wall-to-wall text of difficult material. A busy manager, for example, will want to absorb your information in as easily digestible pieces as possible. Dense text on a page creates noise initially if only because it is so discouraging. If your readers are trying to follow demanding technical information then they are already being challenged, and presenting it to them in solid page-long chunks is at least going to give them mental indigestion. Later, if they want to quickly look up a point you made or a piece of data you presented, they are going to have trouble finding it if they have to wade through a ponderous paragraph until they come to the embedded information. A rule of
>
> *(continued)*

thumb in technical writing states that paragraphs should not be much over twelve lines long, but it's better if they are even less in general. Occasionally you will have to go a little over the twelve-line rule but try not to do so too often. When editing your work, look for any lengthy paragraphs and split them into two. You can always find a place in a long paragraph where you can make a split, even if you have to add a transitional word or phrase. Some of your paragraphs will be much shorter than twelve lines, of course, especially if they are transitional paragraphs or convey particularly complex material. If you are writing a manual or set of procedures, most "paragraphs" will probably be one-sentence directives such as *Move the pointer to the next slide and click again.* One last caution about paragraphs: Try to avoid "orphan lines" in your document—paragraphs for which the first sentence begins on the last line of a page, or the last sentence appears at the top of a page.

Looking at the preceding word mass, you can appreciate a rule of thumb in technical writing: keep paragraphs under twelve lines long, and even less if possible. Occasionally you may have to go over twelve lines, but try not to do so too often. Look for any lengthy paragraphs in your work and try splitting them into two. And when you do, remember that you may have to add a transitional word or phrase.

Some of your paragraphs will be much shorter than twelve lines, of course, especially if they are transitional paragraphs or convey particularly complex material. If you are writing a manual or set of procedures, most "paragraphs" will probably be one-sentence directives such as *Move the pointer to the next slide and click again.*

One last caution about paragraphs: Try to avoid "orphan lines" in your document—paragraphs for which the first sentence begins on the last line of a page, or the last sentence appears at the top of a page.

USE LISTS FOR SOME INFORMATION

A well-organized list is sometimes the most efficient way to communicate information. If you have to present steps in a procedure, materials to be purchased, items to be considered, reasons for a decision—or groceries to be bought—a list might well be the best way to go because readers retrieve some kinds of information from a list more easily than from a passage of prose. Look at the following:

First of all, set the dual power supply to +12 V and −12 V. Next, set up the op-amp, as shown in Figure 1. Use a 1 Vpp sinewave at 1 kHz and then plot the output waveform on the HP digital scope. Then obtain a Bode plot for the gain from 200 Hz to 20 kHz.

This information could be more efficiently presented in list form:

1. *Set the dual power supply to +12 V and −12 V.*
2. *Set up the op-amp, as shown in Figure 1.*
3. *Use a 1 Vpp sinewave at 1 kHz.*
4. *Plot the output waveform on the HP digital scope.*
5. *Obtain a Bode plot for the gain from 200 Hz to 20 kHz.*

There are three main types of lists: **numbered** lists (as above), **checklists**, and **bulleted** lists. You can combine these in various ways to get sublists if you wish. Use a numbered list to indicate when a set of data follows a certain order, as in the preceding example. Numbered lists can also be used to indicate an order of importance in your data, such as a list of priorities or of needed equipment.

Although a **numbered** list implies using numbers, you can substitute upper- or lowercase letters in alphabetical order. Numbers are usually best for the main entries in your list, however, since most people are more familiar with moving through steps 1 through 10 than steps (a) through (k). You can always consider using letters for sublists.

Checklists can be used to indicate that all the items on your list must be tended to, usually in the order presented:

☐ *Connect monitor to computer through the monitor port.*
☐ *Connect keyboard and mouse to computer through the ASF port.*
☐ *Connect power supply to the computer.*
☐ *Connect printer to printer port.*
☐ *Connect modem to modem port.*

When checklists get longer than ten boxes, try to break them down into smaller, more manageable sections and give each section its own subheading.

Bulleted lists are commonly used when items in the list are in no specific order:

Some of the main concerns of environmental engineering are

- *air pollution control*
- *solid waste disposal*
- *public water supply*
- *industrial hygiene*
- *wastewater*
- *hazardous wastes*

Most of today's word-processing software allows you to create bullets easily and to substitute arrows or tick marks if you wish. Lengthy bulleted lists—for example, over seven items—are hard for readers to refer to, so use numbers for longer lists even if no order of priority is intended.

PUNCTUATION AND PARALLELISM IN LISTS

If the lead-in to your list ends with a verb, no colon is necessary. *The five priorities we established are* would not require a colon after *are* since the list is needed to logically and grammatically complete the statement (also see the preceding bulleted list). A lead-in like *We have established the following five priorities* would be followed by a colon, however, since the statement is grammatically complete.

If the items in your list are complete sentences, or contain internal punctuation, you should conclude each one with a period. Otherwise, no punctuation at the end of list items is necessary. Capitalizing the first listed item is up to you, unless each entry is a complete sentence. Whichever style of punctuation and capitalization you use, be consistent.

Another concern when writing lists is to maintain "grammatical parallelism" between entries. This simply means if some entries begin with a verb, all entries should do so; if some begin with a noun, all should. Note how the following list is bumpy due to problems with parallelism:

The accomplishments for WW3-a are as follows:

- *Completion of BIU, ICACHE, and ABUS logic design.*
- *Final simulations on all instruction buffer blocks.*
- *We wrote and debugged 75% of test patterns.*
- *Scheduling of first silicon reticules for WW4-a with Vern Whittington in Fab 16.*

Making the items in the list parallel cuts out some psychological noise:

> *The accomplishments for WW3-a are as follows:*
>
> - *BIU, ICACHE, and ABUS logic design was completed.*
> - *Final simulations on all instruction buffer blocks were run.*
> - *75% of test patterns were written and debugged.*
> - *First silicon reticules for WW4-a were scheduled with Vern Whittington in Fab 16.*

FORMAT YOUR PAGES CAREFULLY

Other factors, besides how you divide information up and how long you make your paragraphs, can have a positive or negative effect on your reader. People prefer print that is visually accessible and pleasing. You can create psychological noise if you fail to meet these preferences, but you can easily prevent it by keeping the following pointers in mind.

MARGINS

Leave ample margins around your text to help prevent your pages from appearing overloaded. Standard margins are one inch all around your page, but it is possible to go a little above or below this if you have to. Make sure the margins are consistent on all pages. If you can, let your lines of text wrap around with a "ragged" right-hand margin rather than aligning them on the right, since this makes for easier reading. If your report is important enough to be bound like a book, you will need a wider-than-usual left margin to accommodate the binding and ensure that the first word or so of each line is still readable.

TYPEFACE

Typeface is the style of individual letters and characters. *Serif* and *sans (= without) serif* are the two general typefaces, with serif fonts having small strokes or stems on the edges of each letter. Books, magazines, and newspapers generally use serif fonts for their text, so this is what people are most used to seeing. Sans serif fonts can be effective for titles and headings, but serif fonts make larger quantities of text more readable since the little stems bind the letters and guide the reader's eye from letter to letter.

sans serif:	The electric car prototype has regenerative braking, which recharges the supply while decelerating the vehicle.
serif:	The electric car prototype has regenerative braking, which recharges the supply while decelerating the vehicle.

Standard type size is 10 to 12 point. You should use larger or smaller sizes only for special effect in titles, captions, warnings, and such. Generally avoid sentences with all capital letters because in a long sequence of uppercase letters you have the same visual outline, making such a sentence difficult to read:

> THE GOVERNMENT PLANS TO ESTABLISH A HIGH-LEVEL ADMINISTRATIVE COUNCIL TO COORDINATE SCIENCE AND TECHNOLOGY.

Capitalized words should be used to emphasize a heading or directive, however:

> DANGER: A 7000 V potential exists across the transformer output terminals.

WHITE SPACE

White space refers to areas of a page not filled with text or graphics. When reading, we tend to take white space for granted but it plays an important part in a document by creating a path for a reader's eyes, isolating and emphasizing important data, and providing "breathing room" between blocks of information. Thus it can have a positive effect by making difficult technical material appear more accessible and less threatening.

You will have enough white space on your pages if you do the following:

- Provide adequate and consistent margins
- Leave a space between all paragraphs
- Leave spaces before and after every heading and subheading
- Leave one or two spaces between text and graphics or lists

- Leave a space before and after each equation in the text
- Indent subheadings or text where appropriate
- Use a ragged (unjustified) right margin

MANAGE YOUR TIME EFFICIENTLY

Few engineers feel they have enough time to do the writing required of them. Often a memo is hastily churned out or a report is rapidly thrown together and tacked on the tail end of a project. As with anything done in a hurry, the results are usually not the best. As the pressure to get a piece of writing out increases, error—that is, noise—also increases. Rather than leaving your writing to the last minute, it is far better to consider it just as much a part of your professional activities as designing, building, and testing.

FINDING AND USING TIME

There are a number of ways to find time to spend on careful writing and editing, but most are not too appealing. You can get to work an hour earlier, or take work home at night (plenty of successful engineers do). You can use your breaks to get away from distractions and concentrate on your writing tasks. You might designate a specific time each day as your writing period—if your colleagues and other duties permit this. You can write on your laptop computer at airports, in flight, on trains, in hotels, or in waiting rooms.

However, as stated above, it's much more effective to make your written work an organic part of your daily schedule. In this way you can assign brief time periods to write short memos and letters or small sections of a report. Larger chunks of time can be designated to concentrate on longer writing tasks.

OUTLINES, DEADLINES, AND TIME LINES

When you have to write anything over two pages long, it's useful to first spend some time making a rough outline. This outline does not have to be set in concrete—that is, you don't have to slavishly follow it once you've written it, and it can be altered at any time—but it will give you some indication of what is involved in producing the finished paper. It will also help you divide your task into smaller sections, which can then be written separately at different times and not necessarily in any order. Less-demanding sections, for example, can be relegated to short periods of available time or to times when more distractions surround you.

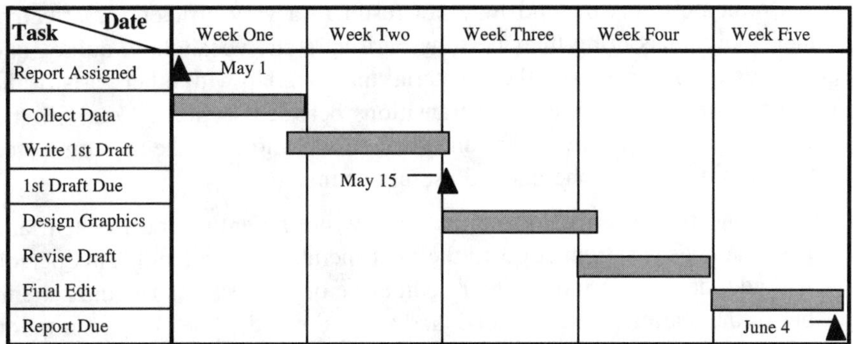

Figure 2-1 The time line you make for your writing project can be as simple or as detailed as you wish. Make sure you have all important tasks and due dates down, however, and then do everything you can to keep to them.

Even if a deadline for completing a document hasn't been imposed on you, it's a good idea to establish one for yourself. Estimate how long you expect the job to take, and schedule back from there. You might even draft out a time line for yourself, showing each date by which you should have completed specific parts of the paper. (See Figure 2-1.) Always allow yourself enough time at the end to review and edit the entire document.

SHARING THE LOAD: COLLABORATIVE WRITING

Not many engineers write lengthy reports by themselves. Technical people work together on project teams for research, design, development, and testing, and often find they must team up to write proposals, manuals, completion reports, and a lot of other technical documents. Team writing is not always easy, especially when people with different degrees of writing ability or ego investment are involved, but if your group plans the task carefully, it can turn out to be relatively painless and very rewarding.

There are three general ways to collaborate on producing a document; some work for some groups, others work for others. We rarely work in ideal circumstances, and you may have to be flexible when working with others on a writing task. The three methods, from the least preferable (but the most commonly used) to the most effective, are as follows:

1. Divide the length of the assignment by the number of people involved, and get each to write his or her share. Individuals will do any research needed for their own section and should write and edit it. Then the document can be "glued together."

Unfortunately this method may not result in a very efficient or effective product. Individuals bring their personal writing style, vocabulary, quirks, and weaknesses to their part, and their material may overlap with other parts of the report or fail to provide important transitions between sections. You will still need a strong writer as "overseer" and final editor who can take the completed draft and mold it into a coherent and useful document.

2. Have one person organize the material, write the entire draft, edit it, and pass the finished product on to the next member of the team. This person will add, delete, rearrange, and re-edit as he or she sees fit. The third member of the team will do the same, and so on down the line. The assumption here is that when all team members have had their say, the document will be as complete and close to perfect as can be.

With a closely knit and cooperative team this method *might* result in an effective report rather than a total mess, but you will need a strong document manager/editor to monitor each step in the process. You might also find this system bringing friendships to an abrupt end. Moreover, if team members want to see what others have done to *their* version of the draft and are inclined to debate and dispute each amendment, you could be a long way past deadline before everything is set right and everyone is satisfied.

3. By far the best way to produce a team document is to assign each member to different tasks according to that member's strengths and interests:
   a. Designate one person as project manager to organize and assign tasks, check that the project is on schedule, and even referee disputes if necessary.
   b. Have another team member get together the needed information for the document, write notes, and put together a very rudimentary draft.
   c. Get the next member, the designated "strong" writer, to generate a working draft of the paper. Ideally, this person is good at writing, enjoys writing—and has read this book.
   d. If possible, get yet another team member with editing skills to act as quality control officer, reading, checking, editing, and in general perfecting the document while working closely with the previous writer.

Using this method, everyone on the team can bring particular strengths to the task and play a significant part in producing the document. Each person has direct access to the document manager, knows what the others' responsibilities are, and has the satisfaction of being uniquely involved in the job. This is the ideal situation. Even with this method you may have to compromise sometimes, double up on tasks, or mix this method with elements of the first two described. Whatever the situation though, carefully planning and assigning collaborative writing tasks to team members before the writing project begins will result in a more efficiently produced document that is both coherent and useful.

EXERCISES

1. Think of some significant communication events you have experienced in the past several months at work or in class. What kinds of audiences were involved? Did a lack of clearly defined audience and purpose cause noise in the communication process? How would a more complete analysis of the audiences have enabled technical information to be communicated more efficiently?
2. Look inside the back cover of an IEEE or other technical journal where you will find a page of advice for authors who wish to publish in that journal. To what extent does the information provide specifications for the articles to be published? Are specifications given for such details as abstracts, length, headings, margins, columns, graphics, size of print, references, and so on? If you still have questions about how a paper for that journal should be written and formatted, look for answers in the Brusaw and Robert Day books listed in this chapter's bibliography or in other standard reference books on technical writing.
3. Go to the library and find a government or industry report on a subject that interests you. Who is the assumed audience? Does the report get to the point right away or does it keep you guessing until the end? How useful are the headings and subheadings? Is it easy to outline the plan of organization the author has used? How do divisions and paragraph length add to the accessibility of the information? Could any of the information be better presented in list form? Select three or four random paragraphs and closely analyze them for ambiguity, wordiness, unnecessary technical jargon, and nouns that could be turned into verbs. Then rewrite those passages.
4. Keep a log of the time you spend writing a document. How long did it take you? Were you working under a deadline? How much time each day was spent planning, writing, and editing? Did others have a part in writing the document, and if so, how were tasks or sections delegated? Were you satisfied with the completed document? Was whoever assigned you the task satisfied with your work? What factors would have enabled you to do an even better job?
5. Take some examples of your own recent writing and analyze them in light of each guideline in this chapter.

BIBLIOGRAPHY

Benson, Philippa J. "Writing Visually: Design Considerations in Technical Publications," *Technical Communication,* vol. 32, no. 4, pp. 35–39, November 1985.

Brusaw, Charles T., Gerald J. Alred, and Walter E. Oliu. *Handbook of Technical Writing,* 4th ed. New York: St. Martin's Press, 1993.

Day, Robert A. *Scientific English: A Guide for Scientists and Other Professionals.* Phoenix, AZ: The Oryx Press, 1992.

Day, Yvonne Lewis. "The Economics of Writing," *IEEE Transactions on Professional Communication,* vol. PC-26, no. 1, pp. 4–8, March 1983.

Markel, Mike. *Writing in the Technical Fields.* New York: IEEE Press, 1994.

3

ELIMINATING SPORADIC NOISE IN WRITING

Problems that crop up intermittently in our writing, often referred to as "faulty mechanics" by English teachers, might be thought of as sporadic or intermittent noise. It tends to occur occasionally, rather than affect every page the way a poor choice of type size or confused organization of material might. Of course enough sporadic noise in a document, such as repeated misspellings or numerous sentence fragments, can easily turn into constant noise and give your reader an impression of hastily and carelessly produced work undeserving of the response or feedback you want.

To help you eliminate sporadic noise, this chapter looks at where it is most likely to occur in spelling, punctuation, sentence structure, and technical usage. We also give some pointers on how to edit your writing in order to remove this kind of noise.

SPELLING AND SPELL CHECKERS

You might think the spell checkers that come with most word processing software have eliminated any need to be a careful speller. Unfortunately this is not the case. With apologies to Shakespeare we took his words "A rose by any other name would smell as sweet," and ran them through a spell checker as *A nose by any outer dame wood small as sweat.* No red flags were raised by the program. Nor will spell checkers catch common errors such as confusing *there* for *their,* or *to* for *too.* Some typographical errors simple give you other words which will pass unnoticed, as in this sentence. A very slight slip of the finger on the keyboard can make the difference between asking for some forms to be mailed to you or nailed to you. A quick transposition could render a memorandum *nuclear* rather than simply *unclear.*

At best, the effect of poor spelling on your readers is a sense of annoyance, or at least of having their attention distracted by something other than what you want to communicate. At worst, noise created by spelling glitches can bring readers to a stop and cause them to seriously question your ability as a writer. They might even suspect that a person who is careless with spelling could also be inept in more critical technical matters.

To reduce or eliminate any noise in your writing caused by incorrect spelling, use a spell checker but also have a standard dictionary nearby. A current dictionary is the only resource that can reliably answer questions such as

- whether there is more than one way to spell a word, or what are the accepted plural forms of words such as *appendix* or *matrix*
- how words like *well-known* or *so-called* are hyphenated, or whether a computer is *on-line* or *online*
- whether it is appropriate to write about *FORTRAN, Fortran,* or *fortran*
- what might be the difference between British and American spelling or usage
- what the accepted past tense is of recent verbs that have come into technical English such as *input*

It is especially important for an engineer to use a current dictionary. English is a dynamic language, and the language of science and technology changes even more rapidly as knowledge increases and devices are developed. You won't find words like *software, modem,* and *LED* in a dictionary from the 1950s, and since then older words such as *bug, hardware, interface,* and even *mouse* have taken on new meanings. Some usage has yet to be decided on: Would a computer shop advertise that it repairs *mice* or *mouses?* Do you send *e-mail, E-mail,* or *email?* (Right now all three options are used.)

PUNCTUATION

Would you want to drive on a busy highway or in a city where there were no traffic signs? Controlling the flow of traffic is vital if anyone is to get anywhere. Similarly, within sentences the flow of meaning is controlled by punctuation marks, the conventionally agreed-upon "traffic signals" of written communication. We do the same thing in spoken language by means of pitch, breath pauses, and emphasis. Directing the flow of ideas in writing is not really difficult, and a useful procedure when you're unsure of how to punctuate a sentence might be to say it aloud as in normal conversation. Pay careful attention to where you pause naturally within the sentence—that is likely to be where you need some punctuation.

Many detailed guides to punctuation exist, and you may want to look at them if you have a lot of queries in this area. You will also find excellent advice on punctuation in the front or back sections of some standard college dictionaries. Meanwhile, the following suggestions are offered on the most common problems many engineers tend to have with punctuation.

COMMAS

Confusion sometimes exists about commas because frequently their use is optional. *Before we arrived at the meeting we had already decided how to vote* would be written with a comma after *meeting* by some and not by others. The question to ask is, Does adding or omitting a comma in a given sentence create noise? In general, if no possible confusion or strain results, the tendency in technical writing is to omit unessential commas.

Often, omitting a comma after introductory words or phrases in a sentence will cause your reader to be momentarily confused—as you would have been if there were no comma after the first word of this sentence. Here are further examples of missing commas causing noise:

> After the construction workers finished eating rats emerged to look for the scraps.
>
> In all the containers were in good condition considering the rough journey.
>
> As you can see the efficiency peaks around 10–12%.
>
> If an acoustic horn has a higher throat impedance within a certain frequency range it will act as a filter in that range which is undesirable.

Solution

> After the construction workers finished eating, rats emerged to look for the scraps.
>
> In all, the containers were in good condition considering the rough journey.
>
> As you can see, the efficiency peaks around 10–12%.
>
> If an acoustic horn has a higher throat impedance within a certain frequency range, it will act as a filter in that range, which is undesirable.

Again, try saying these sentences aloud with their intended meaning. You'll find you put the comma—or pause—where it belongs almost without thinking.

One more point about commas: most technical editors prefer what is called a "serial comma" when you list words or ideas within a sentence, as in *The serial comma has become practically mandatory in most scientific, technical, and legal writing.* You may have been told that the *and* joining the last two terms replaces the need for a comma, but this is not so in technical writing. See how the serial comma is useful in the following sentences by reading them aloud and noting how you need the pause before the *and:*

> Fresnel's equations determine the reflectance, transmittance, phase, and polarization of a light beam at any angle of incidence.
>
> Tomorrow's engineers will have to be able to manage information overload, communicate skillfully, and use a computer as an extension of themselves.

A serial comma may also prevent confusion:

> Rathjens, Technobuild, Johnson and Turblex build the best turbines for our purposes.

Unless *Johnson and Turblex* is the name of one company, you will need a serial comma:

> Rathjens, Technobuild, Johnson, and Turblex build the best turbines for our purposes.

PARENTHESES

Parentheses are used to set off facts or references in your writing—almost like a quick interjection in speech:

> Resistor R5 introduces feedback in the circuit (see Figure 5).
>
> This reference book (published in 1993) contains up-to-date information.

If what you place within parentheses is not a complete sentence, put any required comma or period outside the parentheses:

> Typical indoor levels of radon average 1.5 picocuries per liter (a measure of radioactivity per unit volume of air).
>
> Whenever I design a circuit (like this one), I determine the values of the components in advance.

If your parenthetical material forms a complete sentence, put the period inside the marks:

> I have already calculated the values of the resistors. (R1 is 10.5 KΩ, and R2 is 98 Ω.) The next step is to choose standard values.

Remember, it is best not to use parenthetical material too frequently since these marks force your readers to pause, and are likely to distract them (if only for a brief moment—see what we mean?) from the main intent of your writing.

DASH

A dash will make a sentence seem more emphatic by calling attention to the words set aside or after it: *He was tall, handsome, rich—and stupid.* Since the dash is considered less formal than the other parenthetical punctuation marks (parentheses and commas), you should try to avoid it in very formal writing. If you overuse it, you are in danger of calling wolf too often, and your dashes will lose their effect. With this caution in mind, you may still find dashes helpful for the following purposes:

> *Emphasis:* Staying up all night to finish a lab project is not so terrible—once in a while.
>
> *Summary:* Reading all warnings, wearing safety glasses and hardhats, and avoiding hot materials—all these practices are crucial to sensible workshop procedure.
>
> *Insertion:* My opinion—whether you want to hear it or not—is that the drill does not meet the specifications promised by our supplier.

Notice we are talking about what is called the "em" dash here—the dash used between words that practically touches the letters at each end of it. The "en" dash is shorter, more like a hyphen, and used when you cite ranges of numbers: *31–34; $350–$400.* Most word processing programs allow you to choose whichever you need.

COLON

Colons are used to separate the hour and minute in a time notation and to divide parts of the title of a book or article:

> This proposal is due on Monday morning at 8:30 sharp.
>
> One of the books recommended for the seminar is *The Limits of Safety: Organization, Accidents, and Nuclear Weapons.*

The most common use of the colon within a sentence, however, is to introduce an informal list:

> For the final exam you will need several items: a pencil, a calculator, and three sheets of graph paper.

You can also use a colon to introduce an illustration or example, as we did in the sentence leading into the preceding example. Note, however, that in both cases an independent clause—a statement that can stand by itself and have meaning—comes before the colon. You should **not** write the previous example as

> For the final exam you will need: a pencil, a calculator, and three sheets of graph paper.

because what comes before the colon makes no sense by itself and a colon needlessly interrupts the flow of the sentence. Instead write

> For the final exam you will need a pencil, a calculator, and three sheets of graph paper.

(Note how the same reasoning made us lead into the last two illustrations with no colon after the words ". . . previous example as" and "Instead write. . . .")

HYPHENS

Hyphens have been called the most underused punctuation marks in technical writing. Omitting them can sometimes create real noise, as when we read *coop* (an enclosure for poultry or rabbits) but discover that *co-op* was meant. On the other hand, a hyphen sometimes appears where it is unneeded, as in *re-design* or *sub-question.*

Unfortunately, apart from the general rule that hyphens should be used to divide a word at the end of a line or to join pairs of words acting as a single descriptor, as in *The transistor is a twentieth-century invention*, there is no clear consensus on when to use them. You'll often have to decide for yourself with the help of a recent dictionary, but here are some suggestions:

- Don't hyphenate prefixes such as *pre-, re-, semi-,* and *sub-* unless leaving out a hyphen causes an eyesore or possible confusion. *Preconception* is fine, but *preexisting* needs a hyphen if only for looks. The same might be said of *antiinflationary* or *ultraadaptable.* You may have to distinguish between *recover* (regain) and *re-cover* (to put a new cover on) and the like at times. Again, a good dictionary will help.
- Don't hyphenate compound words before a noun when the first one ends in *ly.* For instance *early warning system* needs no hyphen since it is clear that *early* modifies *warning,* not *system.* The same applies to *optimally achieved goals, highly sensitive cameras,* and similar constructions.
- Stay alert for sentences in which you can eliminate noise by adding one or more hyphens. As you can see, each of the following benefits from the insertion of hyphens at a critical point:

We used a 16 key keypad.
We used a 16-key keypad.

We knew Marienet made klystrons would be able to generate a 9.395 GHz microwave.
We knew Marienet-made klystrons would be able to generate a 9.395 GHz microwave.

The equation assumes a one dimensional plane wave propagation inside the horn.
The equation assumes a one-dimensional plane-wave propagation inside the horn.

Research showed the computer aided students improved their grades dramatically.
Research showed the computer-aided students improved their grades dramatically.

With really complex technical terms you may have very little to go on regarding hyphens. For instance, how do you punctuate *direct axis transient open circuit time constant?* The best solution *(direct-axis transient open-circuit time constant)* may only be found in a technical dictionary or by observing what the common practice is among specialists in the field.

Exclamation Point

The best advice on the exclamation point is to use it all you want in your novel or letters, but avoid it in professional writing except in the case of warnings (DANGER: *Sodium Cyanide is extremely toxic!*). Since engineering documents seek to convey information, any excitement or triumph should be generated by the facts provided in the document rather than by a tagged-on marker. Occasionally an exclamation mark might even be interpreted by your readers as condescending or sarcastic:

> We soon found that the previous data was unsubstantiated!
>
> After reading your report, I feel you might benefit from our on-site course in technical writing!

Quotation Marks

Use quotation marks to set off direct quotations in your text, and put any needed period or comma within them, even if the quoted item is only one word. Although British publishers use different guidelines, the American practice is to always put commas and periods inside quotes, and semicolons and colons outside, as in the following:

> The manager stressed to the whole group that the key word was "Preparedness."
>
> "The correct answer is 18.2 Joules," he told me.
>
> We had heard about the "Four-Star Marketing Plan," but no one remembered what it involved.
>
> We left the game right after the band played "The Eyes of Texas"; it was too hot and humid to stay any longer.

If you have to quote material that takes up more than two lines, set it off from your text by a space and indent it from both right and left margins. You might even use a slightly smaller font size, and should omit the quotation marks, as shown here:

> According to the author, specifications should not be written by a single person:
>
> > The lead engineer delegates the writing of numerous sections to specialists, who may not be aware of the overall goals of the project, and may have parochial views about certain requirements. The lead engineer is faced with the difficult task of fitting all these pieces together, finding all the places where they may conflict, and adjusting them to be correct and consistent with each other [NAWCTSD Technical Report 93-022, p. 11].
>
> The importance of consistency cannot be overstressed in the production of . . . .

SEMICOLON

Whether we like it or not, the semicolon seems to be disappearing from much engineering writing. Often it is replaced by a comma, which is an error according to traditional punctuation rules. More frequently we simply use a period and start a new sentence, but then a psychological closeness might be lost. Look at these two examples:

> Your program is working well, however mine is a disaster.
>
> Take Professor Hixson's class. You'll find he's a great teacher.

The relationship between these statements could be better stressed by using a semicolon:

> Your program is working well; however, mine is a disaster.
>
> Take Professor Hixson's class; you'll find he's a great teacher.

Perhaps one reason we don't see many semicolons in engineering writing is that fewer and fewer people feel confident using them. Another possibility is that little noise results from using a comma or a period and new sentence, as in the preceding examples. Note this pair of sentences:

> We wanted to finish the computer program yesterday; however, the network was down all afternoon.
>
> We wanted to finish the computer program yesterday, however, the network was down all afternoon.

Although the first sentence would be considered correct and the second wrong, you will find plenty of examples of the second punctuation around. The main problem in the second sentence is that a reader can't be sure at first whether *however* "belongs" to the first half of the sentence or the second. A semicolon after *yesterday* is really needed to make this clear. If you frequently use words like *however, therefore, namely, consequently,* and *accordingly* to link what could otherwise be two sentences, insert a semicolon before and a comma after them. You'll find this will add a shade of meaning that cannot be achieved otherwise.

Use semicolons to separate a series of short statements listed in a sentence if any one of the statements contains internal punctuation. The semicolon will then divide the larger elements:

> I suggest you choose one social science subject, such as psychology or philosophy; one natural science course, such as chemistry, physics, or biology; and one math class.
>
> The team is made up of Seth Deleery, vice-president of marketing; Mac Hines, director of research; Ruth Ustby, assistant director of training and human relations; and Leo Little, chief avionics engineer.

SENTENCE SENSE

As engineering writers our aim is to convey information with a minimum of noise. For our purposes the only important "rule" of grammar is to eliminate noise so that the decoder (reader) of the text (message) receives precisely the message the encoder (writer) intends. Thus it's worth looking at those grammatical and stylistic areas where noise often seems to occur in engineering writing. Two persistent but outmoded grammar rules you can safely forget are also discussed in this section under the heading of "Some Latin Legacies."

CONNECTING SUBJECTS TO VERBS

It's unlikely you would write *The machines is broken* without quickly noticing a discrepancy between the subject (*machines*) and the verb (*is*). A problem can occur, however, when several words come between your subject and verb and you forget how you started the sentence. If you are writing in a hurry and leave no time for editing, you might produce something like this:

> This *combination* of electrical components *constitute* a single pole RC filter.
>
> A 35 mm *film* of some high buildings *are* strongly recommended.
>
> Only *one* of the pre-1925 high-rise structures *were* damaged in the quake.

Those plural nouns that follow later (*components, buildings, structures*) can sometimes mislead us into relating the verb to them rather than to the earlier nouns (*combination, film, one*) to which they belong. This danger increases with the length of a sentence and the amount of information intervening between the true subject and verb of a sentence. A good style or grammar program on your word processor may help prevent this from happening, but it is just as well to be alert to the danger.

Sometimes a question arises in engineering writing with units of measurement. Do you write *Twelve ounces of adhesive were added* or *Twelve ounces of adhesive was added?* How about *12 grams of acid was spilled* or *12 grams of acid were spilled?*

The answer is a matter of logic rather than grammar. Even though we're alluding to several ounces or grams here, we "see" them as one unit, and thus the singular verb is preferable. Little or no noise is created, however, if you slip up on this one.

Using *either/or* in a sentence occasionally makes us stop and think. Look at this sentence:

> Either the old manual or the recent procedures (is/are?) acceptable.

Which verb should we use? Since a verb is normally controlled by the noun immediately before it, we would write *Either the old manual or the recent procedures* are *acceptable.* Following this practice we could also write

> Either the recent procedures or the old manual is acceptable.

It is best to follow the same rule with *neither/nor.* Thus the following two sentences would be preferred:

> Neither the engineers nor their supervisor *was* invited to the planning conference.
>
> Neither the rudder nor the wings *were* badly damaged in the crash.

MODIFIERS

A modifier is a word or group of words whose function is to add meaning to other ideas in a sentence. If you say your company has bought a transceiver, you have certainly conveyed some meaning, but if you say *Our company has bought a TS 840 S transceiver with single sideband capabilities,* you add a lot of meaning to the word *transceiver* by adding some modifiers.

The danger lies in creating noise by misplacing the modifiers in a sentence. Such distortion can produce sentences that don't make sense or that make sense in the wrong way. Misplaced modifiers occur when a reader gets the wrong impression (or no impression) of who is doing what in a sentence. This is frequently because words like "I" or "we" or "the engineers" or some other subject has been omitted. Consider the following:

> Jumping briskly into the saddle, the horse galloped across the prairie.
>
> After testing the mechanism, the theory was easily understood.
>
> Once having completed needed modifications and adjustments, the equipment operated correctly and met all specifications.

If we look at these statements logically, we have a horse that rides, a theory that can test a circuit, and equipment that modifies and adjusts. This is not likely to be what the writer meant. Revising the last two sentences might result in the following:

> Once the team had completed needed modifications and adjustments, the equipment operated correctly and met all specifications.
>
> After testing the mechanism we easily understood the theory.

Meanwhile, another problem can crop up if you place a modifier too far from the word or idea it modifies:

> I was ordered to get there as soon as possible by fax.
>
> By the age of four her father knew she would be an engineer.

It's not hard to remedy the lack of logic in these sentences and to avoid traveling by fax or having four year old fathers, but sometimes the meaning cannot be extracted, as in the following:

> The tone-detector circuit was too unreliable to be used in our telephone answering device, which was built of analog devices.

The sentence is correct if the telephone answering device is made of analog devices, but it is much more likely that the writer is concerned with the inaccuracies of an analog tone-detector circuit. This is easily fixed:

> The tone-detector circuit, which was built of analog devices, was too unreliable to be used in our telephone answering device.

PARALLELISM

Parallelism refers to the need for items in a list to share the same grammatical structure. Faulty parallelism creates noise because it violates a sense of logical consistency. Rather than tell someone you *like to jog, wrestling, and play the fiddle,* you would probably say you *like to jog, wrestle, and play the fiddle,* or that you enjoy *jogging, wrestling, and playing the fiddle.* But in longer sentences there is a danger of losing control of this logic.

> After a lot of discussion the team concluded that their alternatives were to call in a consultant, thus increasing the cost of the project, or having three more engineers reassigned to the team.

Note how this sentence reads as if the team's alternatives are (1) to call in a consultant and (2) having more engineers reassigned—two unparallel statements that can grate on our sense of logical flow. The sentence can be rewritten to state that the alternatives were *to call in a consultant . . . or to have three more engineers reassigned.* See if you can recognize the lack of parallelism in this sentence:

> The back-up system should be efficient, should meet safety specifications, and have complete reliability.

To make this statement parallel, think of the list embedded in it. We are told that the back-up system

1. should be efficient
2. should meet safety specs
3. have complete reliability.

To be consistent, the sentence needs one more *should*—or one less:

> The back-up system should be efficient, should meet safety specifications, and should be completely reliable.
>
> The back-up system should be efficient, meet safety specifications, and be completely reliable.

This might seem like a rather fine point, but since a lack of parallelism can often cause a reader to pause, if only subconsciously, it qualifies as noise when it occurs in a sentence. Keeping a parallel structure is even more important when you construct lists, as we pointed out in Chapter 2 under "Use Lists for Some Information."

FRAGMENTS

Sentence fragments are partial statements that create noise because they convey an incomplete unit of information. Here is an example:

> She decided to major in petroleum engineering. Even though it would take five years.

The first sentence makes sense by itself; try saying the second statement alone, as an independent exclamation, and your listeners would be lost. We must admit, however, that in everyday speech and popular journalism you will find plenty of fragments that seem to cause little or no noise. Look at this example:

> Nearly 60% of U.S. households had VCRs by the end of the 1980s. In spite of the microwave oven being the most popular appliance of the decade.

We know what the writer means here, but strictly speaking the second statement is a fragment because it could not stand alone and make sense. The words *In spite of* indicate a contrastive relationship that is clear only in the context of the first statement. It would be more efficient to write

> In spite of the microwave oven being the most popular appliance of the 1980s, nearly 60% of U.S. households had VCRs by the end of the decade.

In your formal engineering writing you would do well to avoid incomplete sentences. They can usually be quite easily remedied, as you can see. Here's another example:

Fragment:	Delays in the October shipments have occurred. Due to the strike.
Complete:	Delays in the October shipments have occurred due to the strike.
Better:	The October 6 shipments have been delayed by the strike.
	or
	The strike has delayed the October 6 shipments.

ACTIVE OR PASSIVE VOICE?

As indicated in the last pair of sentences, we can use two distinct "voices" in English sentences. The active voice directly states that someone does something, as in *The engineer wrote the report.* The passive voice turns it around to *The report was written by the engineer.* Thus the active voice emphasizes the performer of the action—the engineer, in our example—while the passive emphasizes the recipient of the action, the report.

Many engineering and science writers in the past have been advised to leave themselves out of their writing. They might write *It was ascertained that*. . . . long before they would admit *We made sure that* . . . Chances are management would rather tell you *It has been decided to terminate your employment* than *We have decided to fire you.* Perhaps such hedging is necessary at times (it helps conceal responsibility), and there is something to be said for "scientific objectivity," but the natural form of the English sentence is the active voice. This form is also the more efficient one. Look at the following pairs:

Control of the flow is provided by a DJ-12 valve.
A DJ-12 valve controls the flow.

A system for delineating these factors is shown in Figure 5.
Figure 5 shows a system for delineating these factors.

By switching off the motor when it started to vibrate and looking at the tachometer, the resonant frequency was determined.
We determined the resonant frequency by switching off the motor when it started to vibrate and looking at the tachometer.

The use of the passive becomes especially burdensome in procedures or instructions:

The button is pressed twice.

vs.

Press the button twice.

Previously entered data in the database is eliminated by the Edit menu being opened and Select All being chosen.

vs.

Eliminate previously entered data in the database by opening the Edit menu and choosing Select All.

Fortunately, modern engineering writers are getting away from the rigid use of the passive as they realize there is nothing dreadful about using the active voice. Sentences become more vigorous, direct, and efficient in the active form, and by showing that a *person* is involved in the work, you are doing no more than admitting reality. Also, the active voice gives credit where credit is due. If we read in a progress report that *several references were checked out from the library and 25 pages of notes were taken,* are we as impressed by the energy expended as when we read *I checked out several books from the library and took 25 pages of notes?*

One danger of avoiding the active voice is that we can end up writing some pretty absurd things:

> Hurrying to complete the project, several wires got soldered together incorrectly.
>
> The supervisor was seen by us, and we were ignored by her.

Not every use of the passive is inadvisable, of course. Sometimes it will give variety to your writing, and passive verbs can always be used if the doer of an action is unknown or unimportant, or if what is being done is simply more important than the doer:

> Electricity was discovered thousands of years ago.
>
> The bridge was torn down in 1992.
>
> The contaminated material is then taken to a safe environment.

Perhaps the best policy is to use the active voice in your writing if it seems the most natural and efficient way to express yourself, assuming there is no company policy against its use (there is in some companies).

SEXIST LANGUAGE

Gender, or sex, is now only indicated in English by *she/he, his/hers, her/him,* and by a small group of words describing activities formerly pursued by one sex or the other, such as *mailman, stewardess, chairman,* or *seamstress.* Now of course men might bring the drinks on an airplane and women might deliver the mail, not to mention take an equal place in the engineering workplace. Because of this it is unnecessarily restrictive—even offensive—to use gender-specific terms in writing and speech unless there is good reason to do so. The following pairs show how easy it is to reword your sentences or paragraphs to include everyone they should:

> *Restrictive:* Every engineer should be at his workstation by 9 AM.
> *Inclusive:* Every engineer should be at his or her workstation by 9 AM.
>
> or (preferred because less wordy):
>
> Engineers should be at their workstations by 9 AM.
>
> *Restrictive:* An employee can expect a lot of challenges during his career here.
> *Inclusive:* Employees can expect a lot of challenges during their careers here.
>
> *Restrictive:* Every technician must wear safety glasses when he enters the work area.
> *Inclusive:* Technicians must wear safety glasses when entering the work area.

Most nouns indicating gender in English have already been modified to be inclusive. A recent dictionary can guide you here. One title that still sneaks through, however, especially in organizations traditionally dominated by males, is *chairman.* If the "chairman" is female, is she the chairwoman or chairperson? Both are acceptable, but it's probably simpler to refer to anyone in such a position as *chair:*

> Sarah is chair of the new committee on marketing strategy.

or

> Sarah is chairing the new committee on marketing strategy.

SOME LATIN LEGACIES

A few grammar rules impressed upon us in the past really do not hold up under careful linguistic or logical inspection. They were based on how Latin works, rather than English. To put it another way, noise rarely occurs when these rules are ignored. Here are the two main ones, together with comments and a caution.

1. "NEVER END A SENTENCE WITH A PREPOSITION."

In reality a preposition is often the best word to end a sentence with. (A purist might claim we should have just written . . . *the best word with which to end a sentence).* When an editor criticized Sir Winston Churchill for doing so, Churchill responded with "Young man, this is the kind of nonsense up with which I will not put!" After all, did you find any noise in the opening sentence of this paragraph? Efficient writing sometimes dictates that we end a sentence with a preposition. Compare the following pairs. You can see that in each case the first sample, ending with a preposition, flows better and is more natural:

> That's a problem that we will really have to work on.
>
> That's a problem on which we will really have to work.
>
> We must make sure we can find some engineering consultants we can really count on.
>
> We must make sure we can find some engineering consultants on whom we can really count.

2. "NEVER SPLIT AN INFINITIVE."

An infinitive is the form of a verb that combines with the word *to,* as in *to go, to work,* or *to think.* Confident writers have dared *to deliberately split* the infinitive whenever doing so was in the best interests of clear writing. Certain TV space adventurers have been daring *to boldly go* where the rest of us can't for a long time now, and an electrician may find it necessary (and safer) *to entirely separate* the wires in a power line sometimes. But don't overload a split infinitive. If you put too much material between your *to* and the rest of the verb, noise or even nonsense might result:

> The team has been unable to, except for the lead engineer and one technician who is on temporary assignment with us, master the new program.

Rewrite this as

> Except for the lead engineer and one technician on temporary assignment with us, the team has been unable to master the new program.

or

> The team has been unable to master the new program—with the exception of the lead engineer and one technician who is on temporary assignment with us.

TRANSITIONS

Transitional words and phrases are signposts that show a reader the way your thinking is going. They help connect ideas, distinguish conditions or exceptions, or point out new directions of thought. Simple words like *therefore, thus, similarly,* and *unfortunately* eliminate ambiguity by helping a reader interpret your information. So if you neglect transitions in your writing you create noise, since your reader might miss some important connection. Look at these two sentences:

> The group's long-range plans for the S-34B project have been extended. The completion date for the project is as originally planned.

Both sentences are grammatically correct and contain important facts, but can the reader tell how these facts are related? Now notice how the next three illustrations indicate relationships the first example does not:

> The group's long-range plans for the S-34B project have been extended. Nevertheless, the completion date for the project is as originally planned.
>
> The group's long-range plans for the S-34B project have been extended. Unfortunately, the completion date for the project is as originally planned.
>
> Even though the group's long-range plans for the S-34B project have been extended, the completion date for the project is as originally planned.

While facts are important, it is often the relationships between the facts that create the whole picture. Thus you should make your transitions and connections as strong as possible. Here are some examples:

To indicate a sequence: *before...later, first...second, in addition, additionally, then, next, finally—*

Before the project got under way we felt we could never meet the deadline. Later, it became clear there was a realistic chance of doing so.

To indicate contrast: *but, however, yet, still, nevertheless, although, on the contrary, in contrast, on the other hand—*

The GX-40 vehicle scored over 96% in initial dependability testing; nevertheless, the design was scrapped.

To indicate cause and effect: *consequently, therefore, so, thus, hence—*

This company has had to downsize lately. Consequently, many of our staff are looking for other positions.

To indicate elaboration: *further, furthermore, for example, moreover, in fact, indeed, certainly, besides—*

The automotive airbag has proved to be a major factor in driver survival. Moreover, the bag has generated considerable profits for its producers.

SENTENCE LENGTH

When dealing with highly technical subjects you should rarely write sentences over twenty words long. Technical material can be difficult enough to follow without being presented in long, complex sentences, particularly if your audience is less familiar with your field than you are. Even nontechnical ideas are hard to grasp in an unnecessarily longwinded sentence:

We finally had a long discussion with the R & D staff but were not able to convince them that they should commit to a specific date for implementation of the design, but instead they responded with a proposal to extend the project, which would result in a lot more work for all of us and a considerable loss of profits for the company.

Nobody wants to be left breathless at the end of a mile-long sentence. If you find your sentences tend to be lengthy, look for ways to break them into two or more separate ones. The readability of your prose will be determined partly by the length of your sentences. On the other hand, too many short sentences may leave your readers feeling like first graders:

> The Kw766XTR is a low-profile desktop scanner. It has outstanding performance. It offers a frequency range of 29-54 and 108-174 MHz. It includes fifty memory channels. The design is sleek. Individual channels can be locked out. They can also be delayed.

Try to vary your style and avoid both lengthy and abrupt sentences. Remember, however, that very short sentences, used sparingly, can be effective in helping you reinforce a point.

TECHNICAL USAGE

USELESS JARGON

In its negative sense, jargon is pure noise since it refers to unintelligible speech or writing. The word derives from a French verb meaning the twittering of birds, and has a lot in common with "gobbledygook," first used to compare the speech of Washington politicians to the gobbling of Texas turkeys. High-tech jargon is sometimes known as techno-babble. Some people seem to like to sprinkle their writing liberally with such impressive-sounding phrases as *integrated logistical programming, differential heterodyne emission,* or *functional cognitive parameters.* Unfortunately, unless these words hold a precise meaning for both writer and reader, no communication takes place.

Techno-babble is so common that with tongue in cheek we have created an "electrotechnophrase generator" to help addicts satisfy their habit. Select any three-digit number and read off the corresponding words from the following chart; for example 2-8-3 gets *differential heterodyne emission.* Readers may have no idea of what you mean, but they should be impressed—or afraid to ask for a meaning.

The Electrotechnophrase Generator

	Column 1	Column 2	Column 3
0.	voltaic	integrated	simulation
1.	Sholokhov's	semiconductor	algorithm
2.	differential	Yagi	attenuator
3.	Fourier	scaled	emission
4.	transient	Q-factor	diode
5.	virtual	tracking	parameters
6.	phasor	diffusion	network
7.	compound	Doppler	gate
8.	thermal	heterodyne	transducer
9.	Gaussian	coaxial	magnetron

(courtesy of EE students at the University of Texas at Austin)

USEFUL JARGON

In another sense, however, jargon is the necessary technical terminology used in specialized fields. A chemist might use the term *deoxyribose* around a group of peers without feeling a need to explain it, just as a geologist could talk about the Paleozoic era or Devonian period with other geologists. Computer engineers can safely refer to bytes, bauds, and packet switching—among themselves. Communication between experts would be ponderous if not impossible if they had no specialized jargon. Moreover, each year technical language increases greatly as scientific knowledge increases; thousands of technical terms used today were unknown just a few years ago.

Sometimes you will find that common words take on new meanings when used by experts. *Charge, conductor, mole,* and *mouse* are just four examples. Printers (the people, not the machines) mean something quite different than most of us would when they refer to *widows* and *leading*. As engineers, you know and use all sorts of technical jargon. Some you share with most engineers, some with those in the same general field of engineering as you—such as chemical, civil, or aerospace—and some you would use only among peers in highly specialized fields like celestial mechanics or software engineering.

There is only one way to avoid noise when using technical terminology: **know your audience**. Make certain you are writing or speaking at their level of comprehension, because if you're above their heads you will be wasting your time and theirs. Explain terms whenever necessary; don't risk confusing readers or losing them completely because they don't know what you are talking about. Definitions within your text, examples, analogies, or a good glossary are all useful tools for the technical writer who must frequently communicate with less technically inclined audiences. These specific tools are discussed more fully in other sections of this book.

ABBREVIATIONS

Abbreviations are necessary in technical communication for the same reason valid technical jargon is: They refer to concepts that would take a great deal of time to spell out fully. It would be time-consuming and boring for a computer expert to read *Computer-Aided Design/Computer-Aided Manufacturing* several times (or hear it in a talk) when *CAD/CAM* would do. However, you will create a lot of noise in your writing if you use abbreviations your readers don't understand. Always spell abbreviations out the first time you use them unless you know this would insult the intelligence of your audience:

> Then it goes into the ROM (read-only memory).
>
> To understand our billing process, you first need to know what a British thermal unit (Btu) is.

Once you have defined an abbreviation you can normally expect your reader to remember it. The exception to this would be if you are using some highly complicated or unusual abbreviations throughout your document, in which case you may need to remind readers more than once what the abbreviations stand for, or provide a glossary they can refer to.

Initialisms and Acronyms. Abbreviations can be subdivided into **initialisms** and **acronyms.** Initialisms (sometimes called initializations) are formed by taking the first letters from each word of an expression and pronouncing them as initials: GPA, IBM, LED, UHF. Acronyms are also created from the first letters or sounds of several words, but are pronounced as words: AIDS, FORTRAN, NAFTA, NASA, RAM, ROM. Some acronyms become so well-known that they are thought of as ordinary words and written in lowercase characters: *bit, laser, pixel, radar, scuba, sonar.*

Don't be surprised if you find a list of both initialisms and acronyms lumped under the title ACRONYMS. Many engineering writers no longer observe the distinction between the two, and call any abbreviation an acronym. You probably shouldn't make an issue of it, especially if the writer is your superior.

Two usage pointers:

1. Use the correct form of *a/an* before an initialism. No matter what the first letter is, if it is pronounced with an initial vowel sound (for example the letter M is pronounced "em") write *an* before it:

   an MTCR (Missile Technology Control Regime)

   an LED readout

   an SRU pin

   an ultrasonic frequency (*but* a UHF receiver)

   Some abbreviations might fool you. Consider LEM (lunar excursion module) for example. If the custom is to pronounce it as an initialism, L-E-M, then you will have *an* LEM. If it is normally considered an acronym (as one word), you will have *a* LEM.

2. Form the plural of acronyms and initializations by adding a lowercase *s*. Only put an apostrophe between the abbreviation and the *s* if you are indicating a possessive form:

   We ordered three CRTs.

   We were not satisfied with the last CD-ROM's performance.

   or

   We were not satisfied with the performance on the last CD-ROM.

NUMBERS

Engineering means working with numbers a great deal. Frequently this is where a lot of written noise occurs due to typos, incorrect or inexact numbers, and inconsistencies. Obviously you can avoid serious noise by making certain any number you write is accurate. You should also give numbers to the necessary degree of precision: know whether 54.18543 is needed in your report or whether 54.2 will do. Avoid noise from inconsistent use of numbers by following these guidelines:

1. Numbers can be expressed as words (twelve) or numerals (12). Cardinal numbers are one, two, three, etc. Ordinal numbers are first, second, third, etc. Although custom varies, it's a good idea to write the cardinal numbers from one to ten as words and all other numbers as figures.

two transistors	232 stainless steel bolts
three linear actuators	12 capacitors

   However, when more than one number appears in a sentence, write them all the same:

   The IPET has 4000 members and 134 chapters in 6 regions.

Also, use numerals rather than words when citing time, money, or measurements:

1 AM	$5.48	12.4 m	18 ft

2. Spell out ordinal numbers only if they are single words. Write the rest as numerals plus the last two letters of the ordinal:

second harmonic	21st element
fourteenth attempt	73rd cycle

3. If a number begins a sentence, it's a good idea to spell it out regardless of any other rule.

   Thirty-two computers were manufactured today.

   To avoid writing out a large number at the beginning of a sentence, rewrite the sentence so it doesn't begin with a number:

   Last year, 5198 computers were manufactured in this division.

Note You may sometimes see very large numbers written with spaces where you expect commas. Thus 10,354,978 might appear as 10 354 978—to avoid any possible confusion with the practice in some countries of using commas as decimal markers. Decide which method you want to use based on your company's preference and/or your audience.

---

4. Form the plural of a numeral by adding an *s,* with no apostrophe:

80s	1920s

   Make a written number plural by adding *s, es,* or by dropping the *y* and adding *ies:*

nines	sixes
fours	nineties

5. Place a zero before the decimal point for numbers less than one. Omit all trailing zeros unless they are needed to indicate precision.

0.345 cm	12.00 ft
0.5 A	19.40 tons

6. Write fractions as numerals when they are joined by a whole number. Connect the whole number and the fraction by a hyphen:

2-1/2 liters	32-2/3 km

7. Time can be written out when not followed by AM or PM, but you will normally need to be more precise than this. Use numerals to express time in hours and minutes when followed by AM and PM or when recording data. Universal Time (UTC, from the French for *universal coordinated time*) uses the 24-hour clock.

ten o'clock 10:41 AM 8:45 PM

4 hours 36 minutes 12 seconds 23:41 (= 11:41 PM)

8. When expressing very large or small numbers, use scientific notation. Some numbers are easily read when expressed in either standard or scientific form. Choose the best format and be consistent:

0.0538 m or 5.38×10^{-2} m

8.32×10^{-21} m/s 367 345 199 m/s

UNITS OF MEASUREMENT

Although the public in the United States is still not committed to the metric system, you will find that in general the engineering profession is. Two versions of the metric system exist, but the more modern one, the SI (from French *Système International*), is preferred. The vital rule is to be consistent. Don't mix English and metric units unless you are forced to. Be sure to use the commonly accepted abbreviation or symbol for a unit if you do not write out the complete word, and leave a space between the numeral and the unit.

70 ns	100 dB
12 V	34.62 m
23 e/cm^3	6 Wb/m^2

Many people, including technically trained ones, still think in standard or English units of measurement, so sometimes you may find it advisable to give both referents in your writing. As with many other editorial matters, you can only make this decision after thinking of your readers' needs. When it might be advisable to add "explanatory" units, as with a mixed audience, do so by writing them in parentheses after the primary units:

212°F (100°C)	5.08 cm (2 in.)

Make sure you use the correct symbol when referring to units of measurement, and remember that sometimes similar symbols can stand for more than one thing. A great deal of noise would result if you confused the following, for example:

°C (degrees Celsius)	C (coulomb—unit of electric charge)	
g (gram)	G (gauss—measure of magnetic induction)	
m (thousandth)	M (million)	
s (second—as in time)	S (siemens—unit of conductance)	
G (gauss)	G (gravity)	G (giga-)

Units of measurement derived from a person's name usually are not capitalized, even if the abbreviation for the unit is. Note also that although the name can take a plural form, an *s* is not added to the abbreviation to make it plural:

amperes A	farads F	henrys H	
kelvins K	newtons N	volts V	webers Wb

When working with very large or very small units of measurement you will need to be familiar with the designated SI expressions and prefixes:

Factor	**Prefix**	**Symbol**
10^{18}	exa-	E
10^{15}	peta-	P
10^{12}	tera-	T
10^{9}	giga-	G
10^{6}	mega-	M
10^{3}	kilo-	k
10^{2}	hecto-	h
10^{1}	deka-	da
10^{-1}	deci-	d
10^{-2}	centi-	c
10^{-3}	milli-	m
10^{-6}	micro-	m
10^{-9}	nano-	n
10^{-12}	pico-	p
10^{-15}	femto-	f
10^{-18}	atto-	a

A recent dictionary of scientific terms will guide you if you are unsure of the correct spellings or symbols of the units you are using. There is no point in using them in your writing, however, if you or your audience doesn't know what they mean. Symbols and abbreviations are indispensable to an engineer, but use them sparingly when writing for an audience other than your peers. You may sometimes need to define the ones you use, either in your text parenthetically (a brief explanation in parentheses following the term or symbol, like this) or with annotations:

$$P = IE \qquad (1)$$

where
P = power, measured in watts
E = EMF (electromotive force) in volts
I = current in amperes

EQUATIONS

It would be hard to do much engineering without equations. They can communicate ideas far more efficiently than words can at times—consider the ideas represented by $E = mc^2$, for example. However, formulas and equations slow down your reader, so use them only when necessary and when certain your audience can follow them.

Many word-processing programs now make it easy to write equations in text, but if you have to write them in longhand do so with care to ensure both accuracy and legibility. An illegible or ambiguous equation is hardly going to communicate data effectively, and an error in an equation could be disastrous. In other words, make sure your equations are noise free.

You should normally center equations on your page and number them sequentially in parentheses to the right for reference. Leave a space between your text and any equation, and between lines of equations. Also, space on both sides of operators such as =, +, or –. If you have more than one equation in your document, try to keep the equal signs and reference numbers parallel throughout:

$$F(x) = \int \log x \, dx \qquad (1)$$

$$H(s)(xv_2) = X(s)/Y(s) \qquad (2)$$

Eventually you may have to incorporate multiline equations into your technical papers and reports, where they will read (and should be punctuated) just like sentences:

The total harmonic distortion (THD) of voltage at any bus k is defined as

$$THD_k = \frac{\sqrt{\sum_{h=2}^{H} |V_k^h|^2}}{|V_k^1|}. \tag{3}$$

THD can be incorporated into the minimization procedure in [2] by considering a network function that equals the sum of squared THD_k's, or

$$f(I_m) = \sum_{k=1}^{K} (THD_k)^2 = \sum_{k=1}^{K} \left[\frac{\sqrt{\sum_{h=2}^{H} |V_k^h|^2}}{|V_k^1|} \right]^2, \tag{4}$$

$$= \sum_{h=2}^{H} \sum_{k=1}^{K} \frac{1}{|V_k^1|^2} |V_k^h|^2.$$

Note that (4) is identical to (2) when $y(h) = 1$ for $h = 2, 3, 4, \ldots, H$ and when

$$b(k) = \frac{1}{|V_k^1|^2}, k = 1, 2, 3, \ldots, K. \tag{5}$$

Since the fundamental frequency voltages are approximately 1.0 pu, the objective function of (4) is a close approximation to that of (1).

As this example illustrates, no material is too complex to be presented clearly in a flowing, natural manner. Punctuation, transitions, accurate grammar, and mechanics are all indispensable tools for conveying highly technical information with a minimum of noise.

EDIT, EDIT, EDIT

If you look at the early handwritten drafts of some of the greatest writers' works, you'll see alterations, additions, deletions, and other squiggles that indicate how much revision went into the draft before it became a finished work. We could all produce better-written documents if we always

1. **had** the time to edit our work carefully,
2. **took** the trouble to edit our work carefully.

For an engineer, time is frequently going to be a problem. You can't always find time for a leisurely edit of your work. However, you would still be ill-advised to send a first draft of anything of importance to your readers. A quick e-mail note to a friend about lunch isn't worth much concern, but anything more than this, especially if it's going beyond your immediate colleagues, needs at least to be looked over briefly with an editorial eye. How much time you invest in editing should be in direct proportion to the importance of the document. Use all the assistance your word processor will give you, including any spelling, grammar, or readability programs you may have, but don't follow their suggestions blindly. *You* have to be the final arbiter on the clarity and effectiveness of your work—*your* name will be on the document, not your word processor's manufacturer.

EDITING AT DIFFERENT LEVELS

Rather than read over their finished document once or twice in hopes of randomly finding anything in need of improvement, many writers like to take a more methodical approach to editing. You might want to try this. First, check your document for TECHNICAL ACCURACY. Then decide what "writing levels" to approach your editing on, and go through your document at least once on each level. Level 1 might be the highest level of the document, where you check the overall format, organization, and appearance. Is the work arranged the way it should be? Are specifications (if any) followed? Is it the right length? Have you used the best font size, margins, and spacing? Are headings, subheadings, lists, and graphics used effectively?

Level 2 will involve looking at such things as paragraph and sentence length and structure, possible verbiage, and precise word choice. Is the tone of your document appropriate? Have you used the active voice where possible? How about transitions, parallelism, and emphasis where called for?

The final level, Level 3, is the nitty-gritty one of mechanics, spelling, punctuation—all the basics we were supposed to master in elementary and high school. As mentioned above, a good word processing program will provide you with suggestions on spelling and grammar; however, *you* must make the final choices on many of these options.

COLLABORATIVE EDITING

There is nothing wrong with having a colleague, friend, or spouse look over your writing before you submit it to its intended audience. Two heads are usually better than one for discovering flaws in a piece of writing, and you are no longer in a freshman English class where such help might be considered plagiarism. In industry, experts often cooperate in writing technical reports, proposals, and other documents. Most lengthy documents are produced by team effort, where different team members use their particular strengths to ensure that the document is the best it possibly can be—see *Sharing the Load: Collaborative Writing,* in Chapter 2.

Collaborative editing, then, can involve something as simple as asking a friend for his or her opinion of the organization, clarity, and mechanics of your work, and using those comments to improve your writing where necessary. The more skilled and frank your friend is, the better. With a long document, however, collaborative editing can be done by having different team members check the document at different levels, which is usually better than having everyone searching for whatever they can find at all three levels at once.

EXERCISES

1. Review some of your own recent writing for problems with spelling, punctuation, or any of the items listed in this chapter under "Sentence Sense." Did you create any noise in your documents by not following these guidelines? How could you use the guidelines as a "quality control" tool when writing in the future?
2. Find what you feel is a good example of technical writing in any field. Analyze it carefully. What makes it effective, noise-free writing? List and give examples of the ways in which the writer has carefully observed many of the guidelines given in this chapter.
3. Look at an article in any professional journal and determine who its assumed audience is. Then investigate how the author uses technical terminology. Is it appropriate for the audience? Are explanations or definitions given where they seem called for? Do you find any examples of unnecessary technical jargon? How might it have been avoided?
4. Check a number of reports or articles in technical journals that contain abbreviations, numbers, units of measurement, and equations. Are the authors consistent in the way they write these? Does the way these items are written vary from one report to the next, or from one journal to another? In the case of journals, is any information provided on how such things are to be written? Is there any indication in the

journal that a style guide is available for writers who might wish to contribute articles?

5. What are the three levels of editing described in this chapter? Can you think of a document you have written that might have been improved had you used this approach to editing? Find a copy of the document and go through it three times, each time looking for problems on a different level. Give another copy to three friends, asking them to edit it together. Compare what you found as a sole editor with what they find as collaborating editors.

BIBLIOGRAPHY

The Chicago Manual of Style, 14th ed. Chicago: The University of Chicago Press, 1993.

Masse, Roger E. "Theory and Practice of Editing Processes in Technical Communication," *IEEE Transactions on Professional Communication,* vol. PC-28, no. 1, pp. 34–42, March 1985.

Nadziejka, David E. "The Levels of Editing Are Upside Down," *Proceedings of the International Professional Communication Conference,* pp. 89–93, September 28–October 1, 1994.

Rubens, Philip, ed. *Science and Technical Writing: A Manual of Style.* New York: Henry Holt and Company, Inc., 1992.

Weiss, Edmond H. *The Writing System for Engineers and Scientists.* Englewood Cliffs, NJ: Prentice-Hall, Inc., 1982.

4

WRITING LETTERS, MEMORANDA, AND ELECTRONIC MAIL

This chapter focuses on style, format, and content for your professional communications. The first section explores some strategies for deciding which medium of communication to use. The following sections discuss format, style, and content for business letters, memoranda, and e-mail—in that order. The chapter concludes with writing-style issues that apply to communication in any of the media described here and discusses professional communication issues involving the Internet.

WHICH TO USE?

As a working professional, you have at your disposal a variety of communication methods. If you have a question or request to make, you can run down the hall and issue it in person; if it's to someone external to your organization, you can write a business letter; if it's to someone within your organization, you can write a memo; if it's urgent or informal, you can make a telephone call; and, if you have a properly equipped computer, you can send e-mail.

LETTER OR MEMO?

Memoranda are written communications that stay within an organization (a business firm or a government agency, for example). Business letters are written communications to recipients who are external to the organization of the sender. Of course, some internal communications are in the form of business letters—for example, those letters that the CEO sends out once or twice a year to all employees.

VAPOR OR PAPER?

The decision whether to use telephone or face-to-face communication as opposed to some form of written communication is fairly clear. In telephone or face-to-face communication, there are these issues:

- *Permanent record.* There is no record of what transpires in your phone conversation.
- *Availability of the recipient.* Recipients of the communication may not be in their offices or at their desks, forcing you to play "telephone tag" for a frustrating period of time.
- *Attitude of recipients.* Recipients may not take the communication as seriously as they would if it were in writing.
- *Purpose, length, and complexity of the topic.* Some topics are just too much for a conversation. For example, it would be difficult to present details of product specifications or a proposal over the phone or face to face.

BITS OR BLOTS?

The decision gets harder when you must choose between e-mail and a printed medium such as a memorandum or a letter. Once you get used to e-mail, you may wonder why you should bother with phone calls, business letters, or memos at all. After all, e-mail eliminates worrying with stamps and envelopes and finding a mailbox—not to mention the delay in delivery and response times. Better than telephone communication, e-mail doesn't require its recipients to be in the right place at the right time—they can read it when they are ready. And, unlike telephone communication, there is a record of the communication. However, there are certainly instances where print remains the preferable, and sometimes the only medium of communication:

- *Recipients.* Obviously, if some of the recipients don't have e-mail, then printed letters or memos are necessary.

- *Contents of the message.* If the message contains material that cannot be conveyed through ordinary e-mail—such as graphics or tables—then the printed letter or memo is necessary.
- *Need for reply or forwarding.* If the letter or memo contains pages that the recipient must fill out and send, printed copy may be preferable even though e-mail certainly supports such reply and forward functions. You can't expect every member of the organization to be a whiz at less common functions of e-mail like these.
- *Security issues.* As Ed Krol points out in *The Whole Internet User's Guide and Catalog,* you can assume that any e-mail you send has a chance of being seen by anyone in the world. While that's stretching it a good deal, there is an element of truth to it. You may not want to send confidential information (new product specifications or sensitive data about a project or colleague) by e-mail.
- *In-person discussion of the memo.* If the message must be used in face-to-face situations of any sort, the printed memo may be necessary. If everybody's going to have to print out the memo before coming to the meeting, you might as well print it for them and thus eliminate one more snag.
- *Importance or length of the information.* For some, e-mail lacks the feeling of settled, established information. It's light, ephemeral stuff—just not a medium for serious business. The nature and contents of your communication may not be right for electronic mail. People seem to be less likely to study, concentrate on, and take seriously an electronic message displayed on a computer screen than they are a memorandum or letter printed out and lying on their desk.
- *In-your-face factor.* For some people, a printed memo sitting on their desk just cannot be avoided. Of course, that depends—for some professionals, it's more inconvenient to fetch hardcopy mail than to check e-mail. Still, people who get their e-mail through commercial services such as America Online, CompuServe, Delphi, or Prodigy cannot stay logged in all day. At $0.03 to $0.10 a minute, the bill mounts up fast. Ultimately, you have to base your decision on which medium your colleagues are most in the habit of using.

BUSINESS LETTERS

As mentioned previously, the common business letter (printed on real paper!) is not dead. Face-to-face, telephone, and e-mail communications are just not right for certain kinds of correspondence. Use a hardcopy letter when you want to make sure that the recipient receives it and takes it seriously, when you want the recipient to be able to study it at length and act appropriately upon it, when the communication is long and packed with information, or when you want to ensure that there is a permanent record of the communication. Use the following format and design suggestions for business letters—professional communications external to your organization.

STANDARD COMPONENTS OF BUSINESS LETTERS

The following describes standard components for business letters, most of which are illustrated in Figure 4-1. Of course, not all these components occur in any individual letter.

- *Company or personal logo.* If you use company stationery, begin your letter a full inch below the logo. Don't use logo stationery on following pages; there should be matching stationery without the logo. If you are an independent consultant, design your own logo! With all the tools now available in advanced word-processing software, you can create a logo with a larger, fancy type style, maybe some combination of bold and italics, and maybe horizontal lines above or below your name, title, and address.
- *Heading.* The heading portion of the letter contains the sender's address and the date. If you're using letterhead stationery, only the date is needed.
- *Inside address.* This portion of the letter includes the name, title, company, and full address of the recipient of the letter. Make this the same as it appears on the envelope. This element is important if the secretarial staff opens your letter and discards the envelope.
- *Subject line.* Some business-letter styles make use of a subject line, the same kind that you see in memoranda. You use this element to announce the topic, purpose, or both of the letter—for example, "Request for copyright status on the X11 documentation" or "In response to your request for copyright status."
- *Salutation.* This is the "Dear Sir" element of the letter. In some contexts where there is no clear recipient or the recipient does not matter, the salutation is omitted. If you feel you must include a salutation but don't know how to address it, there are several possibilities. Call the recipient's organization and attempt to get a specific name (as well as title and department name). Create a department or group name that is reasonably close. For example, if you are writing a recommendation letter, write "Dear Recruitment Officers:" Of course, if all else fails, you can use the infamous TO WHOM IT MAY CONCERN:. Notice that the salutation for business letters is punctuated with a colon. (A comma implies a friendly, nonbusiness communication.)

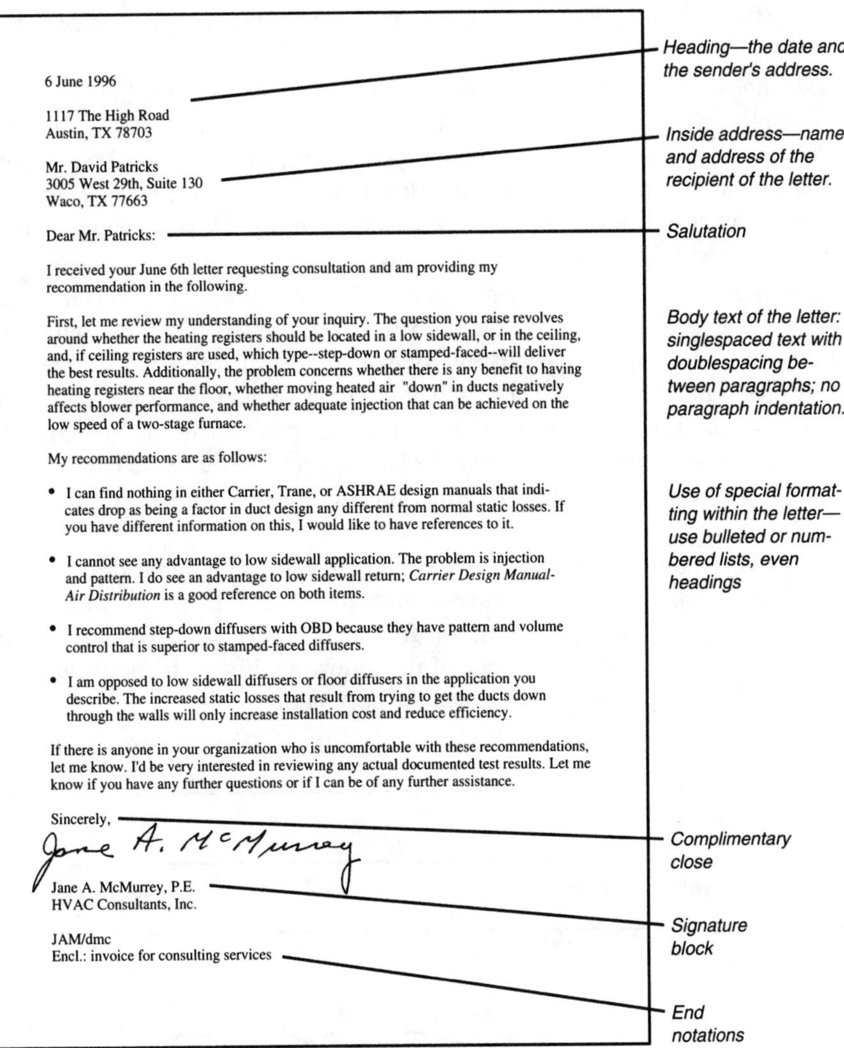

6 June 1996

1117 The High Road
Austin, TX 78703

Mr. David Patricks
3005 West 29th, Suite 130
Waco, TX 77663

Dear Mr. Patricks:

I received your June 6th letter requesting consultation and am providing my recommendation in the following.

First, let me review my understanding of your inquiry. The question you raise revolves around whether the heating registers should be located in a low sidewall, or in the ceiling, and, if ceiling registers are used, which type--step-down or stamped-faced--will deliver the best results. Additionally, the problem concerns whether there is any benefit to having heating registers near the floor, whether moving heated air "down" in ducts negatively affects blower performance, and whether adequate injection that can be achieved on the low speed of a two-stage furnace.

My recommendations are as follows:

- I can find nothing in either Carrier, Trane, or ASHRAE design manuals that indicates drop as being a factor in duct design any different from normal static losses. If you have different information on this, I would like to have references to it.
- I cannot see any advantage to low sidewall application. The problem is injection and pattern. I do see an advantage to low sidewall return; *Carrier Design Manual-Air Distribution* is a good reference on both items.
- I recommend step-down diffusers with OBD because they have pattern and volume control that is superior to stamped-faced diffusers.
- I am opposed to low sidewall diffusers or floor diffusers in the application you describe. The increased static losses that result from trying to get the ducts down through the walls will only increase installation cost and reduce efficiency.

If there is anyone in your organization who is uncomfortable with these recommendations, let me know. I'd be very interested in reviewing any actual documented test results. Let me know if you have any further questions or if I can be of any further assistance.

Sincerely,

Jane A. McMurrey

Jane A. McMurrey, P.E.
HVAC Consultants, Inc.

JAM/dmc
Encl.: invoice for consulting services

Figure 4-1 Standard business-letter formats—the block letter.

- *Body of the letter.* The body of the letter begins just after the salutation and continues until the complimentary close. Typically, the text of business letters is singlespaced; the first line of paragraphs is not indented; and doublespacing is used between paragraphs.
- *Complimentary close.* This is the "Sincerely yours," element of the letter. In letters where interpersonal action is irrelevant, it too is sometimes omitted. If there is more than one word in the complimentary close, capitalize

only the first word. Notice that the complimentary close is punctuated with a comma.

- *Signature block.* This element is variously defined as everything from the complimentary close through and including the end notations to just the blank area for the signature and the typed name. In professional correspondence, don't forget to include those letters after your name that identify the degree or title that you worked so hard to earn. Below your name, include your title and the name of your company.
- *End notations.* These elements are the "Cc:" and "Encl:" abbreviations that you often see below the sender's typed name. The first set is the initials of the sender and typist, respectively (for example, "JMC/rbs"). "Cc:" followed by one or more names indicates to whom a copy of the letter is sent. Abbreviations like "Encl." or written-out versions such as "Enclosure" or "Attachments" indicate that other documents have been attached to the letter. If you want, you can specify exactly what you've attached: for example, "Encl.: resume."
- *Following pages.* As mentioned earlier, if you use letterhead stationery, use the matching pages (the same quality and style of paper but without the letterhead) on following pages. On following pages in professional correspondence, use a header like one of the following, in which you include the name of the addressee, the date, and the page number:

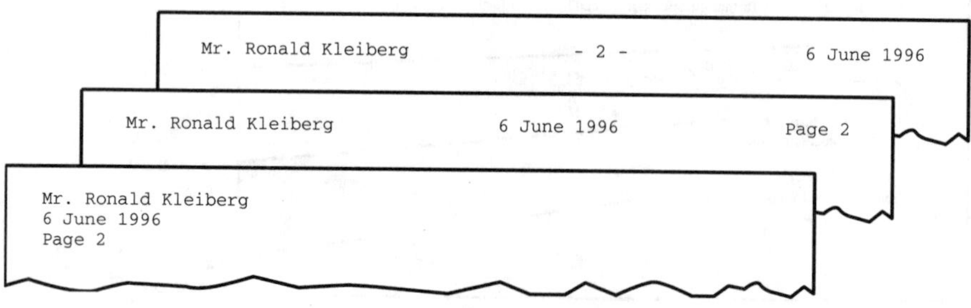

COMMON BUSINESS-LETTER FORMATS

Traditionally, business letters have used one of four standard formats: the *block,* the *semiblock,* the *alternative block,* and, more recently, the *simplified* formats. These formats vary strictly according to which elements are present (for example, is there a salutation?) and where they are placed on the page (for example, is the heading on the left or right margin?).

Figures 4-1, 4-2, and 4-3 show letters formatted for each of these designs.

Clarkson Hall, Rm. 1709
Monash University
Clarkson, WA 98881
25 May 1996

Hughes, Gano, Associates
1118 The High Road
Austin, TX 78703

Dear Colleague:

I am writing to professional consultants like yourself in attempt to survey any experience you have had using different dynamic solvers to solve undamped linear eigenproblems, particularly large eigenproblems (with greater than 5000 degrees of freedom).

There is a wide variety of solvers available. They vary from subspace iteration, to Lanczos, to conjugate-gradient, to dynamic condensation, and to component mode synthesis. I would like to know what professional engineers are using and why. (For example, is your choice faster, more robust, or is there some other reason?)

At present, my interests are focused on solving unsymmetric linear equations in the boundary element method. However, from a practical viewpoint, I have attempted to solve a liquid oscillation problem by using "pseudo-fluid" elements in NISA. The trouble involves choosing a Poisson's ratio as close to 0.5 as possible.

I found the Lanczos method to be best in this case, but there were other difficulties in simulating the boundary conditions at the top of the fluid with appropriate springs. For these reasons, I ultimately had to abandon this idea. Even so, the subspace and accelerated subspace iteration techniques were not nearly as effective.

I am currently doing research in the area of accelerating the solution of linear undamped eigenproblems, and I'm interested in comparing what the actual users find is most useful (and not just the theoretical researchers!).

I would very much appreciate hearing about any experience or insights you may have had in these areas. If it would be easier for you, you can contact me by e-mail; my address is janemc@pink.cc.monash.edu.us.

Sincerely,

Jane A. McMurrey

Jane A. McMurrey, P.E., Ph.D.
Monash University

Figure 4-2 Standard business-letter formats—the semiblock letter.

- *Block format*—The most commonly used. It's the easiest; all the elements are flush left.
- *Semiblock format*—Similar to the block format except that the heading, complimentary close, and signature block are on the right margin.
- *Alternative block format*—Adds a subject line; otherwise, it uses the same design as the semiblock format.
- *Simplified format*—Omits the salutation; otherwise, it uses the same design as the block format.

For communications that involve no professional interaction, the simplified and the alternative formats are acceptable. Notice that the letter for the job announcement in Figure 4-3 uses the alternative format (or could have used the simplified format). But for serious professional communications, such as proposals or employment letters, stick with the block or semiblock format.

REPORT-LIKE LETTERS AND COVER LETTERS

If you are writing a report or some other type of "standalone" document, you'll want to attach a cover letter to the front when you send it. This is a brief business letter that announces what the report is about, why it was written, for whom, and other such identifying details.

You can incorporate a report into the framework of a business letter. You can present an engineering report, complete with tables, illustration, lists, and headings—all within the confines of the common business letter.

BUSINESS MEMORANDA

You can use a memorandum for most communications internal to an organization. It might be a call to employees for a general meeting; it might be a reminder that status reports are due; it might contain a status report; it might be a request to an employee to provide information; it might be that employee's subsequent report. Thus a memo can be very much like a business letter, or it can be very much like a short report—the key is the memorandum format.

STANDARD COMPONENTS OF MEMORANDA

Memoranda use a format that is much simpler than that of business letters. Figures 4-4 and 4-5 illustrate the standard components.

25 May 1996

Dr. Patrick H. McMurrey
Department of Mechanical Engineering
Clarkson Hall, Rm. 1709--Monash University
Clarkson, WA 98881

SUBJ.: Position for experienced development engineer

Dear Colleague:

CSMI is a leading sawmill equipment manufacturer headquartered in Portland, with manufacturing facilities in Portland and Hot Springs, AR.

We are looking for a seasoned (8 to 10 years) development engineer with a hands-on style and a strong background of stress analysis and design optimization for large capital equipment. Bachelor's degree in mechanical engineering is required; advanced degree preferred.

CSMI offers competitive compensation, company-paid health, dental, life and pension. Optional 401(k). CSMI is a drug-free workplace. We are also an equal opportunity employer; qualified applicants who would enhance our cultural diversity are encouraged to apply.

To be considered, please submit resume with salary history and requirements to:

Human Resources Manager
CSMI
4000 NW St. Helens Rd.
Portland, OR 97210

Figure 4-3 Standard business-letter formats—the alternative block letter. The simplified letter format omits the complimentary close.

- *Date: heading.* While formats vary, the date you send the memo should be somewhere in the header. The example in Fig. 4-4 shows it as the third line in the header; in some designs it is the first line, as in Fig. 4-5.
- *TO: heading.* Put the name of the recipient or the group name in the "TO:" heading. The level of formality is very apparent here. You can put "Sarah" or "Bob," "Sarah James," or "Ms. Sarah James, Director of Personnel," depending on your familiarity with the recipient and the formality of the situation.
- *FROM: heading.* Put your own name or the name of the person or group for whom you are writing the memo in this slot. Once again, your familiarity with the recipient and your sense of the formality of the situation dictate whether to put just your first name, your full name, or your full name and title. In many organizations, the writer of the memo jots his or her initials or first name just after the typed or printed name.
- *Subject heading.* In this slot, you place a brief phrase that encapsulates the topic and purpose of the memo. For example, if you have surveyed the capabilities of grammar-checking software, your subject might be "Results of our survey on grammar-checking software." The actual label for this element varies: some styles use "RE:" or "SUBJ.:".

  If your memo is in response to something, phrase the subject line accordingly. For example: "Re your request for a grammar-checker survey" or "Review of your grammar-checker survey results."
- *Signature block.* In more formal styles of memoranda, writers actually insert the same kind of complimentary close and signature block that you see in business letters.

REPORT-LIKE MEMOS AND COVER MEMOS

When you write a short report, for example, under four pages, you have two choices as to the format. You can put the entire report into the memo—headings, lists, graphics, the works. Or you may prefer to create a cover memo and attach the report as a separate document. The cover memo briefly announces the topic and purpose of the attached information, provides an overview of its contents, and some request for review or response. For example, see Figures 4-4 and 4-5.

ELECTRONIC MAIL

Once you get comfortable with e-mail, it may seem like the only communication method you need for your professional work. However, written media are

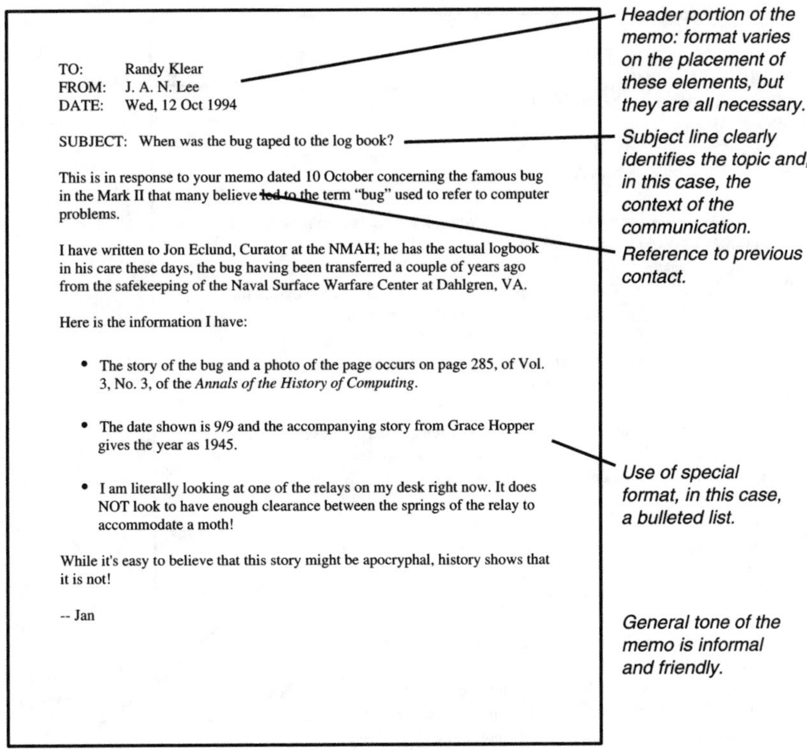
TO: Randy Klear
FROM: J. A. N. Lee
DATE: Wed, 12 Oct 1994

SUBJECT: When was the bug taped to the log book?

This is in response to your memo dated 10 October concerning the famous bug in the Mark II that many believe led to the term "bug" used to refer to computer problems.

I have written to Jon Eclund, Curator at the NMAH; he has the actual logbook in his care these days, the bug having been transferred a couple of years ago from the safekeeping of the Naval Surface Warfare Center at Dahlgren, VA.

Here is the information I have:

- The story of the bug and a photo of the page occurs on page 285, of Vol. 3, No. 3, of the *Annals of the History of Computing*.
- The date shown is 9/9 and the accompanying story from Grace Hopper gives the year as 1945.
- I am literally looking at one of the relays on my desk right now. It does NOT look to have enough clearance between the springs of the relay to accommodate a moth!

While it's easy to believe that this story might be apocryphal, history shows that it is not!

-- Jan

Header portion of the memo: format varies on the placement of these elements, but they are all necessary.
Subject line clearly identifies the topic and, in this case, the context of the communication.
Reference to previous contact.
Use of special format, in this case, a bulleted list.
General tone of the memo is informal and friendly.

Figure 4-4 Example of a business memorandum.

preferable in a number of ways, as discussed at the beginning of this chapter. The following focuses on e-mail functions, style, and format.

IMPORTANT E-MAIL FUNCTIONS

If you are just starting to use e-mail, you should master several skills right away. Many different e-mail programs are out there, and unfortunately you may have to learn more than one (see the examples of different e-mail systems in Figure 4-6). Whichever system you use, make sure you master the following functions:

- Sending e-mail
- Editing e-mail before sending it
- Sending e-mail to multiple addresses
- Receiving and reading e-mail
- Saving e-mail into files or folders
- Replying to e-mail and forwarding e-mail
- Creating and using aliases and distribution lists

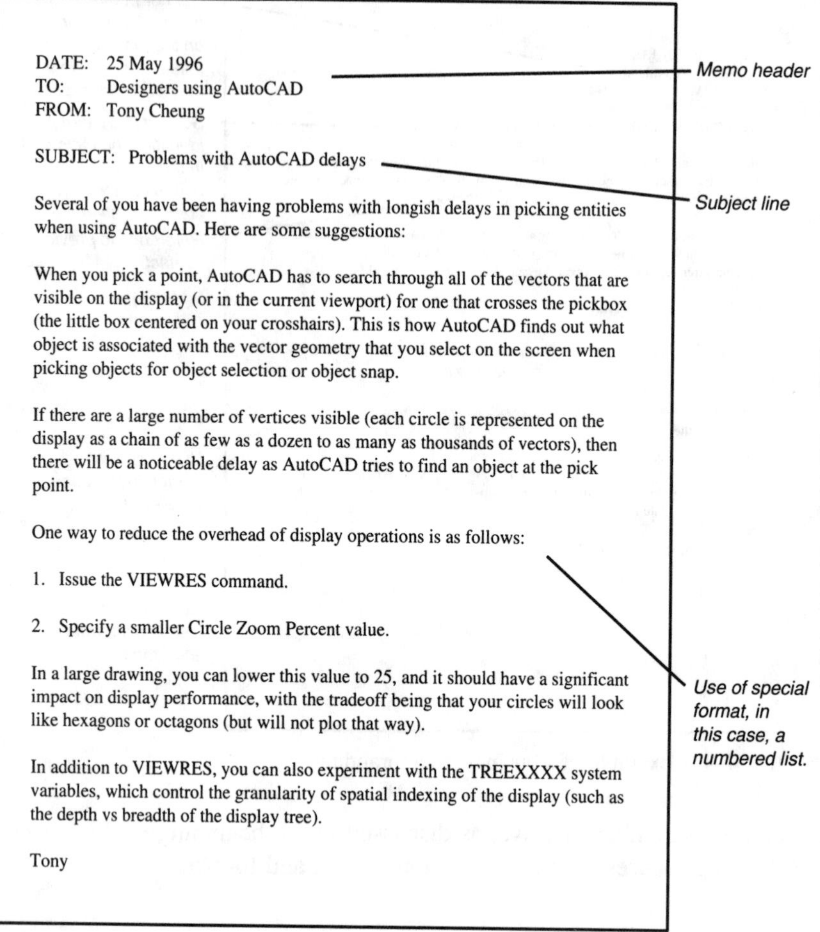

DATE: 25 May 1996
TO: Designers using AutoCAD
FROM: Tony Cheung

SUBJECT: Problems with AutoCAD delays

Several of you have been having problems with longish delays in picking entities when using AutoCAD. Here are some suggestions:

When you pick a point, AutoCAD has to search through all of the vectors that are visible on the display (or in the current viewport) for one that crosses the pickbox (the little box centered on your crosshairs). This is how AutoCAD finds out what object is associated with the vector geometry that you select on the screen when picking objects for object selection or object snap.

If there are a large number of vertices visible (each circle is represented on the display as a chain of as few as a dozen to as many as thousands of vectors), then there will be a noticeable delay as AutoCAD tries to find an object at the pick point.

One way to reduce the overhead of display operations is as follows:

1. Issue the VIEWRES command.

2. Specify a smaller Circle Zoom Percent value.

In a large drawing, you can lower this value to 25, and it should have a significant impact on display performance, with the tradeoff being that your circles will look like hexagons or octagons (but will not plot that way).

In addition to VIEWRES, you can also experiment with the TREEXXXX system variables, which control the granularity of spatial indexing of the display (such as the depth vs breadth of the display tree).

Tony

Figure 4-5 Example of a business memorandum.

- Attaching ASCII and "binary" files to e-mail
- Keeping copies of the e-mail you send

Practically every e-mail program handles these functions differently and even refers to them by different names. Whatever the implementation, it's well worth the half-hour it takes to learn these functions so that you can run your professional work smoothly and effectively.

E-MAIL FORMAT AND STYLE

Format for the heading portion of e-mail messages is quite simple, most of it being handled by the e-mail program itself. You fill out the address (or

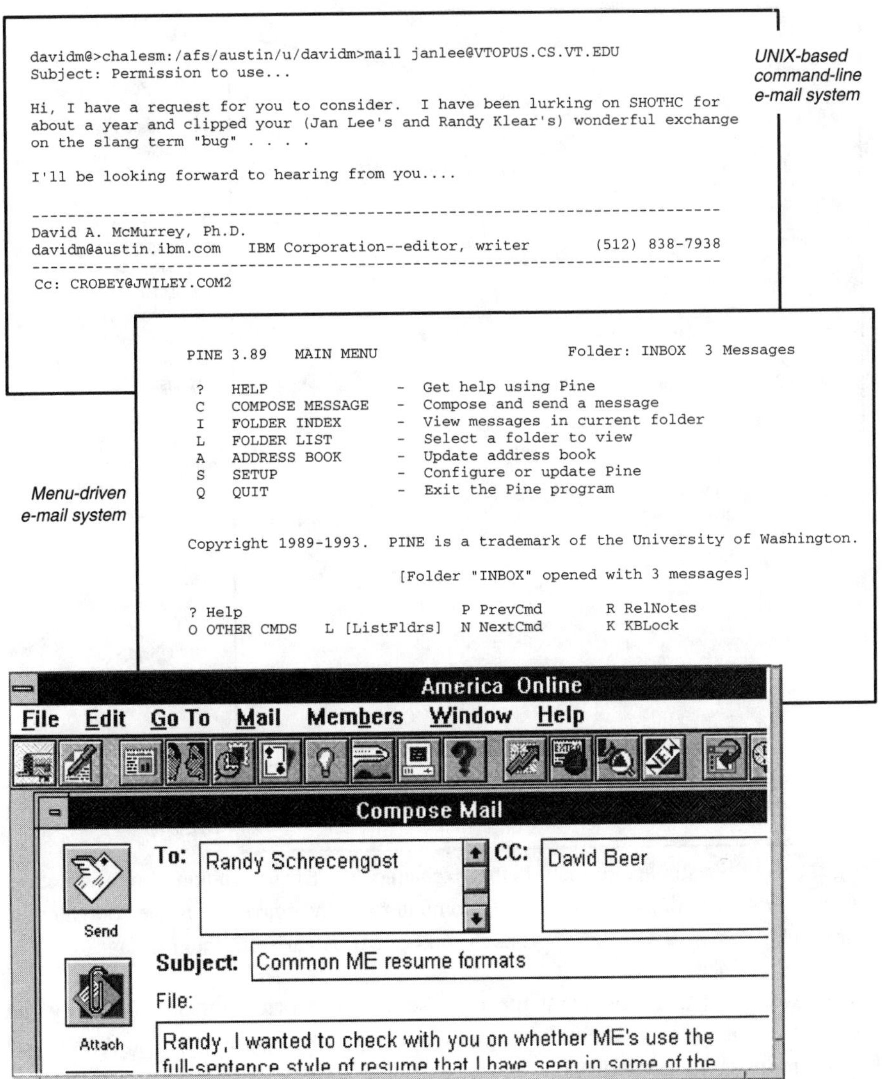

Figure 4-6 Examples of e-mail programs. Command-based e-mail programs require you to know commands and subcommands. Menu-driven systems enable you to choose from lists of options, making life easier. Windows-based systems enable you to click on the actions you need to perform, which makes things easier.

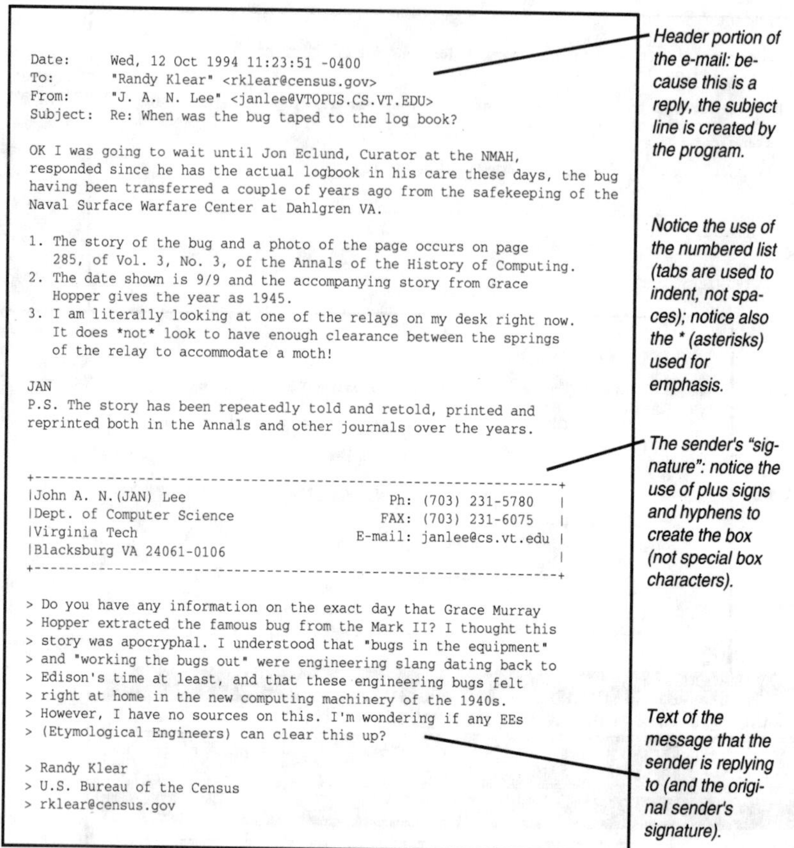

```
Date:     Wed, 12 Oct 1994 11:23:51 -0400
To:       "Randy Klear" <rklear@census.gov>
From:     "J. A. N. Lee" <janlee@VTOPUS.CS.VT.EDU>
Subject:  Re: When was the bug taped to the log book?

OK I was going to wait until Jon Eclund, Curator at the NMAH,
responded since he has the actual logbook in his care these days, the bug
having been transferred a couple of years ago from the safekeeping of the
Naval Surface Warfare Center at Dahlgren VA.

1. The story of the bug and a photo of the page occurs on page
   285, of Vol. 3, No. 3, of the Annals of the History of Computing.
2. The date shown is 9/9 and the accompanying story from Grace
   Hopper gives the year as 1945.
3. I am literally looking at one of the relays on my desk right now.
   It does *not* look to have enough clearance between the springs
   of the relay to accommodate a moth!

JAN
P.S. The story has been repeatedly told and retold, printed and
reprinted both in the Annals and other journals over the years.


+-------------------------------------------------------------+
|John A. N.(JAN) Lee                       Ph: (703) 231-5780 |
|Dept. of Computer Science                FAX: (703) 231-6075 |
|Virginia Tech                   E-mail: janlee@cs.vt.edu     |
|Blacksburg VA 24061-0106                                     |
+-------------------------------------------------------------+

> Do you have any information on the exact day that Grace Murray
> Hopper extracted the famous bug from the Mark II? I thought this
> story was apocryphal. I understood that "bugs in the equipment"
> and "working the bugs out" were engineering slang dating back to
> Edison's time at least, and that these engineering bugs felt
> right at home in the new computing machinery of the 1940s.
> However, I have no sources on this. I'm wondering if any EEs
> (Etymological Engineers) can clear this up?

> Randy Klear
> U.S. Bureau of the Census
> rklear@census.gov
```

Header portion of the e-mail: because this is a reply, the subject line is created by the program.
Notice the use of the numbered list (tabs are used to indent, not spaces); notice also the * (asterisks) used for emphasis.
The sender's "signature": notice the use of plus signs and hyphens to create the box (not special box characters).
Text of the message that the sender is replying to (and the original sender's signature).

Figure 4-7 Example of e-mail. In this exchange, Mr. Klear is looking for information on the famous bug that got into an early computer and thus gave rise to the computer term. In his response, Dr. Lee appends the text of Mr. Klear's original message

addresses) to which you want the mail sent, add a brief phrase indicating the contents or purpose of the message, and specify any other address to which you want the message copied (Cc:).

An additional formatting element to add to your professional e-mail communications is a "signature." Instead of just signing your name at the bottom of your note, construct a signature containing your full name, title, company you're with, e-mail addresses, regular mailing address, telephone number, and other such detail. Signatures work almost like business cards; many people use them to get in a quick bit of self-advertisement. You can see these elements in the example e-mail message shown in Figure 4-7.

As for style in e-mail messages, here are some suggestions:

- *Typos and other kinds of mistakes.* There is some controversy about how much to worry about writing mistakes in e-mail. While no one wants to send out messages with typos or grammar problems, the speed at which we typically generate e-mail makes it inevitable. This problem is compounded by the fact that some e-mail software makes it difficult to go back and edit preceding lines. Except for very formal electronic communications, most people disregard or even expect occasional writing glitches in the e-mail they send or receive.

- *Informality.* The tone of e-mail communications is generally informal. At its best, electronic communications is a wonderful, ongoing, international group-help conference with everybody pitching in and either asking for help or providing it.
- *Brevity.* E-mail messages are normally rather short—for example, under a dozen lines—and the paragraphs are short as well. No one likes having to do a lot of extended reading on a computer screen. When they must send a lengthy note, e-mail senders often put a warning toward the top of the message or even in the subject line such as "Lengthy note follows..."
- *Specific subject lines.* If you want your e-mail to be read and have the impact you intend, make the subject line specific and compelling. Once people get involved with e-mail, it's not uncommon for them to log in and find sixty to seventy messages waiting. All that mail appears in a simple list with only the senders' computer IDs and subject lines showing. Therefore, if you want your message to get recognized for what it contains, the subject line must convey as much as possible within the space of some 25 to 35 characters.
- *Put the most important information at the top of the message.* Active e-mail users tend to lose interest or patience quickly. Make sure that the most important information gets in at the beginning of your message. Some of us tend to beat around the bush before we get down to our main topic, point, request, or whatever it is we are writing about. On the Internet, we'll also tend to get ignored!
- *Limit the width of your messages to sixty characters.* For some recipients, the width of your message is not a problem; their e-mail software will reformat it. But for others, it can be a real headache: lines that are too long break in odd, distracting ways. For that reason, limit your line to sixty characters.
- *Use short paragraphs and skip an extra line between paragraphs.* Whenever possible, break your messages into paragraphs of less than six or seven lines. And when you divide your message into paragraphs, skip an extra line between them.
- *Take a different approach to emphasis.* In e-mail communications, we don't have underscores, bold, italics, different type styles or sizes—at least not yet. Some e-mail writers resort to all-caps. Maybe an occasional NOT is okay, but extensive use of capitalization for emphasis is unpleasant to read.

Instead, adopt the e-mail world's approach to emphasis: *not* or _not_ or <not>.

- *Use only the basic character set.* Use only the characters that you see on your keyboard: the letters, the numbers, and the symbols that you see on the keytops. Anything else may not format properly when the recipient views it. For example, you might be tempted to use the ASCII box characters to set up a border around your e-mail signature. For some recipients, that won't work! Use the - (hyphen) and the | (bar) and the + (plus sign) keys only, as shown in Figure 4-7.
- *Use headings and lists.* If your e-mail message is long (say, over forty lines), use headings to identify and mark off the various subtopics within the message. If you have key points to emphasize, if there is a series of points, or if there is step-by-step information to send, use the various forms of lists that are available. Use an * (asterisk) in bulleted list items. Use a simple number followed by a period for numbered list items.
- *Use tabs, not spaces, for indentation and other special alignment and margins.* In word-processing programs like WordPerfect, AmiPro, and Word, you may have noticed that you can't create an indented paragraph by hitting the spacebar the same number of times for each indented line. The same is true for most e-mail messages. Use the tab key instead.
- *Exercise caution with humor and sarcasm.* Skillfully working humor and sarcasm into writing is hard in any medium, but it's particularly challenging in electronic communication. Typically, we write e-mail messages rapidly and send them off without as much scrutiny and revision as we would a print-based communication. This makes humor and irony a chancier affair. And the risk is increased because recipients generally read their e-mail much more rapidly than they do printed copy. Many people end up using "smileys" to ensure that their humor, sarcasm, irony, or other attitude is interpreted correctly on the receiving end. (See the examples on the following page.)
- *Be careful with automatic replies.* The reply function in e-mail is a wonderful time-saver. It eliminates the business of having to create a new message and type in the recipient's address. However, e-mail is often addressed to multiple recipients. Imagine that a colleague has written e-mail to you to discuss some aspect of a project and that, in your reply, you raise some questions about the competence of another partner in the project. But what if your colleague had copied (Cc:) that partner on his original note? Uh oh... Always check the original message for other recipients.

```
:) or :-) Smile—typically used to indicate a sarcas-
tic, joking, or witty statement.

;) or ;-) Wink—another way of indicating wit, irony,
sarcasm, humor.

:( or :-( Sad face—indicates a kind of mock sadness.

*:o) Clown face—indicates that the writer is "clown-
ing around."

:| or :-|Indifferent, bored, or glum face.
```

WRITING STYLE IN BUSINESS CORRESPONDENCE

Regardless of the medium in which you conduct your business correspondence, most of the guidelines for writing style are the same. Thus, whether you are writing a business letter, memorandum, or e-mail message, the following recommendations are equally valid.

- *Indicate the topic in the first sentence, if necessary.* If your subject line hasn't made it clear already, state the topic and purpose of your communication in the very first sentence. Don't force readers to wade through lots of background to get to the point.
- *Identify any preceding correspondence or situation to which your communication is a response.* In the first paragraph, establish the context of the communication by referring to whatever previous meeting, phone conversation, or correspondence has taken place.
- *Provide an overview of the contents of the communication if necessary.* If the letter, memo, or e-mail is lengthy, provide a bit of overview of the contents. This needn't be more than an informal list in a sentence within the first paragraph.
- *Keep the paragraphs short.* Ideally, paragraphs in business correspondence should not go over six to eight lines. In business correspondence, readers are less willing to wade through long, dense paragraphs than they are, for example, in textbooks or formal reports.
- *Consolidate the contents of paragraphs.* In attempting to keep paragraphs short, however, don't start new paragraphs just anywhere. Divide your communication into paragraphs at those points where the topic or focus changes.

- *Use headings for communications over a page in length.* If your communication is more than a page or two and if the information in it is like that in a report, use headings to mark off the boundaries where new topics start and stop.
- *Use lists and graphics as you would in a report.* Business correspondence can at times resemble reports; writers use the same sorts of headings, lists, graphics in their letters and memos. Look for lists of items, particularly in the longer paragraphs, and try turning them into vertical lists. Similarly, work graphics (illustrations, diagrams, tables) into your correspondence just as you would in regular reports.
- *Be brief, succinct, to the point.* The importance of brevity is never so great as it is in business correspondence—and still more so in e-mail communications. Readers in this context don't have much patience with unnecessary introduction and background or wordy ways of expressing ideas.
- *In memos and e-mail, interact with your readers; be as informal as the situation allows.* Because memoranda are internal to an organization and usually addressed to people in your area, your writing style can be informal. Whenever appropriate, use the "you" style of writing—avoid the impersonal third-person and passive-voice styles.
- *Indicate any action necessary on the part of the recipient.* Let readers know what you expect them to do as a result of reading your correspondence. What actions should they take after reading your letter, memo, or e-mail? Fill out a questionnaire? Where is it located? Where should they send it? Make sure that all details like these are clearly and specifically explained.

PROFESSIONAL COMMUNICATIONS ON THE INTERNET

Until now, professional interchange and networking occurred mostly through associations, conferences, seminars, journals and newsletters, business letters, and, of course, telephone conferences and face-to-face meetings. The Internet offers additional options. For example, some conferences are held online, which saves the expense of travel, food, and lodgings and cuts down on the cost of registration. Some professional journals are available online; you can subscribe to the online version rather than wait for the hardcopy version to come in the mail. Whole courses of study, books, and reference materials are now available online. And of course, much professional correspondence takes place online.

INTERNET RESOURCES

The Internet also provides some additional modes of professional communication that are rather different from anything we are used to:

Electronic Mailing Lists, Newsletters, and Newsgroups. Electronic mailing lists and newsgroups permit ongoing discussions of professional topics. Typically, these mailing lists and newsgroups are dedicated to specific topics or specific areas of professional activity. For example, civil engineers and manufacturing engineers have newsgroups dedicated to their professions. In these newsgroups, you see lots of e-mail loaded with questions, answers, debates, job advertisements, and all sorts of helpful information. Here are some examples of electronic mailing lists:

Adv-Eli Discussion of latest advances in electrical engineering sponsored by a Chilean-based IEEE student group.

Adv-Elo Discussion of latest advances in electronics sponsored by a Chilean-based IEEE student group.

Aelflow Discussion of aerospace and aeronautical engineering.

CAEDS-L Discussion of computer-aided engineering design products.

Energy and Climate Information Exchange (ECIX) Newsletter Newsletter focusing on energy and climate issues.

ChE Electronic Newsletter An electronic newsletter for chemical engineers.

EEJobs Postings and discussion of positions for electrical and electronic engineers.

CIRCUITS-L Discussion of circuit analysis for electrical engineering undergraduates.

IEEE-L Discussion that acts as forum for all IEEE student branch officers and members.

CIVIL-L Discussion of civil engineering and education.

MECH-L Discussion of mechanical engineering topics.

Anonymous FTP (File-Transfer Protocol) Facilities. The file-transfer facilities of computers on the Internet enable professional engineers to establish archives of reports, graphics, standards, specifications—databases of all sorts. FTP is the Internet's standardized mechanism for moving files from one computer on the Internet to another. *Anonymous FTP* refers to the ability to log in to a computer without a special ID or password and transfer files from it to your own computer. Through anonymous FTP, you can go out to these archives, browse around, and copy materials that you need. Or perhaps a colleague will make materials available to you in FTP archives.

World Wide Web Facilities. The web provides an even more exciting set of possibilities for your professional life. Internet facilities such as e-mail, electronic mailing lists and bulletin boards, and anonymous FTP are each separate activities that demand their own learning curve. The World Wide Web consolidates most of these activities into a unified interface. You can join mailing lists, read the communications taking place in a newsgroup, send e-mail, browse FTP archives, read journals, scan conference proceedings, rummage through specifications and standards, look for services and colleagues—all using the same interface. Many professionals are now "setting up shop" on the web. They set up a "home page" with their picture and a resume of their professional services and work history. This works much like a Yellow Pages entry in the telephone book, except that the potential audience is global.

GETTING STARTED ON THE INTERNET

If you're an absolute, ground-zero beginner on the Internet, here are some suggestions:

- Get a modem and communications software so that you can connect your computer to the Internet. If your university, company, or other organization already has an Internet connection—lucky you! Just ask for access (and use the magic words, "business purposes").
- Start an account with one of the Internet service providers such as Delphi, America Online, Prodigy. But check locally—there may be local service providers right in your home town. Whichever service you get, demand the following:
  - *Full e-mail*—Make sure you can correspond not just with other members of the same service but with anybody anywhere on the Internet (which means anywhere in the world).
  - *File-transfer*—Make sure that you can transfer files from other Internet computers to your space on your service's computer as well as transfer files from your space to other sites. (This is FTP.)
  - *Download and upload*—Confirm that you can move files from your own local computer (the one sitting on your desk) to your space on your service provider's computer, and vice versa.
  - *Telnet capabilities*—Ensure that you can log in to another computer on the Internet if you have a valid login name and password for that computer; this is called "telnet." Occasionally, you'll need to log in to another computer half-way around the world and do some work there—all from your own office.

- *Work space*—Make sure that you can store files in your own designated work space on your service's computer and be able to work with them there. Watch out, though. The more you store and the longer you store it, the more you pay.
- *Chat mode*—Perhaps not as important as these other functions—at least not yet—is the ability to have "real-time" conversations online. Sometimes, e-mail is too slow; you need the immediacy of a telephone conversation (but for some reason, phone connection is not possible or practical). In chat mode, you type a statement and press the Enter key; your chat partner reads it and responds to it with a statement. Chat-mode sessions roll by on your computer screen looking very much like the script of a play.
- *Usenet access*—We've already discussed those Internet newsgroups with quirky names such as the following:

```
alt.cad.autocad        alt.comp.hardware.homebuilt
alt.hvac               comp.unix.security
comp.software.eng      sci.engr.manufacturing
sci.engr.civil         sci.techniques.spectroscopy
sci.electronics        sci.engr.robotics
```

  These are referred to collectively as *Usenet newsgroups;* there are well over 5000. While some view these newsgroups as the exclusive province of Elvis sighters and DOOM maniacs, serious professional interchange does occur there, as the names of the newsgroups previously mentioned show. Make sure the service you use gives you access to the Usenet newsgroups; find out if there are any limitations on the number of newsgroups that are provided.
- *World Wide Web access*—You should demand the capability to use the "web." Make sure that you can set up a "home page" containing whatever you want the world to know about yourself.
- *Other stuff*—The big commercial online services offer lots of "stuff" that can be bewildering when you're shopping for a service: for example, stock market quotes, games, cookbook recipes, forums, software libraries, selections from news magazines, as well as sports and weather. These are fun and may have professional importance to you, but for the most part they are distractions. They make it harder to figure out whether a service offers the functions you really need.

EXERCISES

Schedule interviews with at least three professional engineers, and ask them questions about the business correspondence they receive:

1. What are the typical audience, purpose, and content of their letters and memos? Why do they write a letter or memo instead of just making a phone call?
2. How much secretarial assistance do they receive? Do they get any help editing or proofreading their correspondence?
3. When they have to convey specialized, technical information, is it to another engineer, or do they have to do a lot of translating of technical detail for nonspecialists?
4. How do they decide between writing a hardcopy letter or memo, making a phone call, or sending e-mail?
5. What do they see as the advantages of using e-mail in conducting their business? What problems do they see in using e-mail as a business communications tool?

BIBLIOGRAPHY

Angell, David, and Brent Heslop. *The Elements of E-Mail Style.* Reading, MA: Addison-Wesley, 1994.

December, John, and Neil Randall. *The World Wide Web Unleashed.* Indianapolis, IN: Sams, 1994.

Gilster, Paul. *The Internet Navigator.* New York, NY: Wiley, 1994.

Krol, Ed. *The Whole Internet: User's Guide and Catalog.* Sebastapol, CA: O'Reilly, 1994.

Maxwell, Christine, et al. *New Riders' Official Internet Yellow Pages.* Indianapolis, IN: New Riders Publishing, 1994.

The Internet Unleashed. Indianapolis, IN: Sams, 1994.

5

WRITING SOME COMMON ENGINEERING DOCUMENTS

This chapter explores some of the common types of reports you may write as an engineer, particularly in terms of their typical content and organization. As you read this chapter, keep in mind that the names of these types vary considerably, and their contents often combine in different ways. You may want to skim the following list to find what you need in this chapter:

Inspection or trip report: Briefly reports on the inspection of a site, facility, or property; summarizes a business trip; or reports on an accident, describing the problem, discussing the causes and effects, and explaining how it can be avoided.

Laboratory report: Reports on an experiment, test, or survey; presents the data collected, discusses the research theory, method, or procedure; discusses conclusions, and, possibly, explores applications of the findings or possibilities for further research.

Specifications: Provide detailed requirements for a product needed by an organization or detailed descriptions of existing products; provides specifics on design, function, operation, and construction.

Progress report: Summarizes how your project is going, what you or your group has accomplished, what work lies ahead, what resources have been used, what problems have arisen.

Proposal: Seeks a contract, approval, or funding to do a project; functions as a competitive bid to get hired to do a project; promotes yourself or your organization as a candidate for a project; promotes the project itself, showing why it is needed.

Instructions: Explains to employees or customers how to perform certain tasks, provides procedures on using certain equipment, gives troubleshooting and maintenance guidelines, explains policies and operating procedures.

Recommendation report: Studies a situation or problem, reports on various alternatives or options, recommends the best one or assesses the feasibility of an idea.

Note The reports discussed in this chapter are mostly short and informal and are routinely formatted as in-office memoranda or business letters. However, practically any of these reports can be formatted as full-length formal reports.

SOME PRELIMINARIES

Before going to any of the main sections of this chapter, consider a few points that apply to all the types of reports about to be discussed:

- *Don't get hung up on the names of reports.* Sorry, there is no ANSI standards committee on the proper names, contents, and format of reports. Don't worry about whether a report is really an evaluation report or a recommendation report. Ask as many questions as you can about the requirements of the specific report you must write; think as clearly as you can about the needs and requirements of the audience of the report. One or some combination of the types of reports discussed in this chapter is likely to suffice.
- *Think creatively about content and organization.* Rarely will these suggestions on report contents and organization be a perfect fit for your real-world report projects. The plans for reports presented here really cannot be used as templates. They are more like a library of subroutines in a programming project—they are all out there as tools in case you need them. Always be thinking creatively, brainstorming about what else your readers may need and what else the report-writing situation calls for.
- *Build your reports on the needs of your audience.* Everything about your report depends on the specific people who are going to read it. Remember that sometimes you must write for different audiences within the same report. See Chapter 2 for detailed discussion of analyzing and adapting to audiences.
- *Be careful with discussion of technical background.* Most of these types of reports *may* require technical background—but not necessarily. Background sections are supposed to provide information to make the rest of the report

understandable. However, poorly written background sections are so loosely related that they are not helpful. To avoid this common problem, review the main text of your report and list what readers may need help with; only then write the background section, and with those needs squarely in mind.

- *Be careful with the report introduction.* Another problem concerns introductions. Remember that the introduction introduces readers to the report, *not* the technical subject matter. The introduction gets the reader ready to read the report. It announces the topic, alludes to the situation that brought about the need for the report, indicates what the audience needs to know in order to understand the report, and provides a brief overview of the topics to be covered (or not covered). Other elements may be necessary, given the nature of the specific report. But it's bad practice to dive right into the main subject matter—readers then lack any perspective, overview, or roadmap for the whole report.
- *Find out what your company's requirements are.* This chapter illustrates common format, contents, and organization for reports. Remember, however, that every company, organization, field, or profession has its own names for reports as well as its own requirements on format, content, and organization. Those requirements may be written down in some official document, or they may be traditions that everybody "just knows." Your job is to find out what those requirements or traditions are. The discussion and examples in this chapter give you some clues about what to expect, and something to use when there are no guidelines.

INSPECTION AND TRIP REPORTS

One common group of engineering reports handles tasks such as reporting on the inspection of a site or facility; describing an incident or accident, with details on the causes, effects, and ways to prevent; summarizing events and results of a business trip; describing property, equipment, or new technology. You might hear these types of reports referred to variously as site reports, inspection reports, incident reports, trip reports, or accident reports.

These reports are similar in that they contain lots of description, narration, discussion of related causes and effects, as well as a certain amount of interpretation and evaluation. If you were reporting on an accident, you'd describe the damage, then explore the causes. If you were reporting on a business trip, you'd narrate the events; if the purpose of the trip had been to assess a new technology, you'd also do some evaluation. Obviously, these types of short reports overlap considerably. It's a loose category—the names are by no means fixed or widely agreed upon.

- *Trip report:* Discusses the events, findings, and other aspects of a business trip. This type of report documents your experience and observations so that people in your organization can share it (see Figure 5-1).
- *Investigation or accident report:* Describes your findings concerning a problem; explores its causes, its consequences, and what can be done to avoid it.
- *Inspection or site report:* Reports your observations of a facility, a property, or an installation of equipment, with both description and possibly evaluation of it.

If the computer age has changed anything about how professionals write, it's in this area of short, informal reports. Many reports like these are now being composed and transmitted strictly as e-mail. For example, if you had been to a conference on new surface mount techniques, you'd want to make your discoveries available to colleagues back at work. Upon your return from the conference, you could send everyone e-mail or, better still, put your report in some commonly accessible electronic location.

However, with e-mail, you lose graphics, special fonts, highlighting, italics, bold, underlining, and different type sizes. Giving up special features such as these is one of the trade-offs you must make if you want to go strictly electronic.

CONTENTS AND ORGANIZATION

For the content of your informal report, consider these suggestions:

Introduction. No matter which type of report you write, it's a good idea to begin by indicating the purpose of the report and providing a brief overview of its contents. Avoid that impulse to dive right into the thick of the discussion!

Background. It's also a good idea to set the stage—to explain the background or context of the report. Why did you go on this business trip? Why were you sent to inspect the facility? Who sent you? What are the basic facts of the situation—the time, date, place, and so on.

Factual Discussion. The main contents of a report like this are in part description or narration. Typically, you must describe the accident, facility, property, or the proposed equipment. Give a blow-by-blow account of what happened on the trip: where you went; who you met with; what the discussions were about.

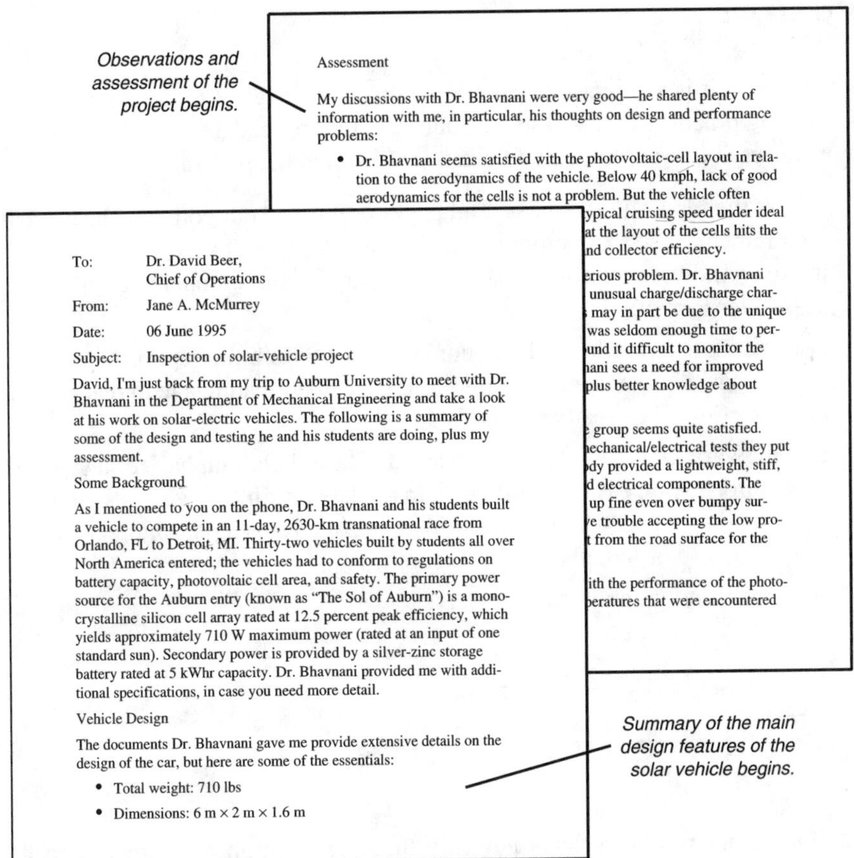

Observations and assessment of the project begins.
Summary of the main design features of the solar vehicle begins.

Assessment

My discussions with Dr. Bhavnani were very good—he shared plenty of information with me, in particular, his thoughts on design and performance problems:

- Dr. Bhavnani seems satisfied with the photovoltaic-cell layout in relation to the aerodynamics of the vehicle. Below 40 kmph, lack of good aerodynamics for the cells is not a problem. But the vehicle often ...ypical cruising speed under ideal ...at the layout of the cells hits the ...nd collector efficiency.

...rious problem. Dr. Bhavnani ... unusual charge/discharge char... may in part be due to the unique ... was seldom enough time to per... und it difficult to monitor the ...ani sees a need for improved ... plus better knowledge about

... group seems quite satisfied. ...echanical/electrical tests they put ...dy provided a lightweight, stiff, ...d electrical components. The ... up fine even over bumpy sur...e trouble accepting the low pro... t from the road surface for the

...ith the performance of the photo...eratures that were encountered

To: Dr. David Beer, Chief of Operations

From: Jane A. McMurrey

Date: 06 June 1995

Subject: Inspection of solar-vehicle project

David, I'm just back from my trip to Auburn University to meet with Dr. Bhavnani in the Department of Mechanical Engineering and take a look at his work on solar-electric vehicles. The following is a summary of some of the design and testing he and his students are doing, plus my assessment.

Some Background

As I mentioned to you on the phone, Dr. Bhavnani and his students built a vehicle to compete in an 11-day, 2630-km transnational race from Orlando, FL to Detroit, MI. Thirty-two vehicles built by students all over North America entered; the vehicles had to conform to regulations on battery capacity, photovoltaic cell area, and safety. The primary power source for the Auburn entry (known as "The Sol of Auburn") is a monocrystalline silicon cell array rated at 12.5 percent peak efficiency, which yields approximately 710 W maximum power (rated at an input of one standard sun). Secondary power is provided by a silver-zinc storage battery rated at 5 kWhr capacity. Dr. Bhavnani provided me with additional specifications, in case you need more detail.

Vehicle Design

The documents Dr. Bhavnani gave me provide extensive details on the design of the car, but here are some of the essentials:

- Total weight: 710 lbs
- Dimensions: 6 m × 2 m × 1.6 m

Figure 5-1 Excerpts from a short business-trip report. The writer summarizes her visit with researchers involved in the solar-vehicle design and provides an assessment of that work.[1]

Actions Taken. If you are investigating a problem and implementing a solution, your report should contain a step-by-step discussion of how you determined the problem and corrected it.

Interpretive, Evaluative, or Advisory Discussion. Once you've laid the foundation with the background and factual discussion, you're set to do what is probably expected—to evaluate the property or equipment, explain what caused the accident, interpret the findings, recommend further action, or recommend ways to prevent the problem in the future.

[1] Information for this report was developed from S. H. Bhavnani, "Design and Construction of a Solar-Electric Vehicle," *Journal of Solar Energy Engineering* (February 1994), 28–34.

FORMAT

Just because it is a short and informal report, don't neglect to use simple, basic formatting practices that will make your report more readable, more usable, and more accessible—not to mention more professional in appearance.

- Unless the report goes over several pages or unless your company has certain requirements, use the memorandum format. (And obviously, if you are writing to an individual or organization not a part of your company, use the business-letter format.)
- Use headings to mark off the major subtopics within the report. Notice how they are used in the example report in Figure 5-1. Headings help readers skip to the sections they really want to read.
- Use the various types of lists as needed. These help emphasize key points, make information easier to follow, help readers return to key points, and generally create more white space which makes your report more readable. Notice the use of bulleted lists in Figure 5-1.
- Use tables and graphics as necessary.

LABORATORY REPORTS

The laboratory report presents not only the data from an experiment and the conclusions that can be drawn from that data, but also the theory, method, procedure, and equipment used in that experiment. As much as practical, the laboratory report should enable readers to "replicate" the experiment so that they can verify the results for themselves. Because of this dual requirement, laboratory reports have a characteristic structure.

CONTENTS AND ORGANIZATION OF LABORATORY REPORTS

The term *laboratory report* is actually a bit restrictive. *Any* report in which you gather original data through an experiment, field research, or survey (for example, an opinion poll) and then draw conclusions from those findings can use the content and organization plan discussed in the following. (See the excerpts from a laboratory report in Figure 5-2.)

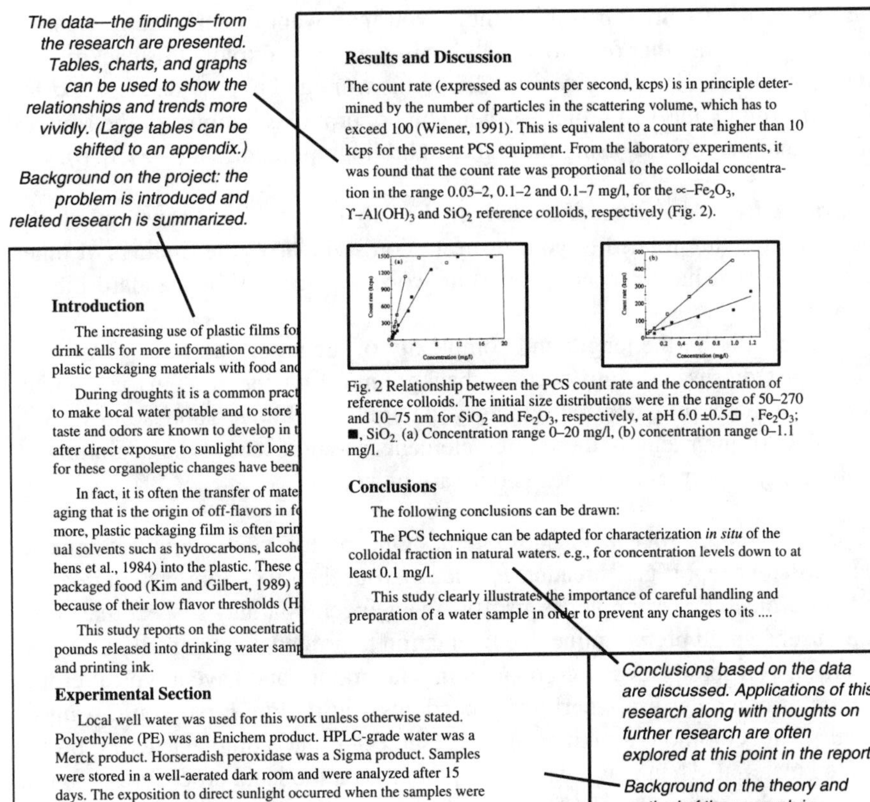

The data—the findings—from the research are presented. Tables, charts, and graphs can be used to show the relationships and trends more vividly. (Large tables can be shifted to an appendix.)
Background on the project: the problem is introduced and related research is summarized.
Conclusions based on the data are discussed. Applications of this research along with thoughts on further research are often explored at this point in the report.
Background on the theory and method of the research is discussed; procedures and facilities are described.

Introduction

The increasing use of plastic films fo drink calls for more information concerni plastic packaging materials with food and

During droughts it is a common pract to make local water potable and to store i taste and odors are known to develop in t after direct exposure to sunlight for long for these organoleptic changes have been

In fact, it is often the transfer of mate aging that is the origin of off-flavors in f more, plastic packaging film is often prin ual solvents such as hydrocarbons, alcoh hens et al., 1984) into the plastic. These c packaged food (Kim and Gilbert, 1989) a because of their low flavor thresholds (H

This study reports on the concentratio pounds released into drinking water samp and printing ink.

Experimental Section

Local well water was used for this work unless otherwise stated. Polyethylene (PE) was an Enichem product. HPLC-grade water was a Merck product. Horseradish peroxidase was a Sigma product. Samples were stored in a well-aerated dark room and were analyzed after 15 days. The exposition to direct sunlight occurred when the samples were put on the roof of the building for 15 days in June.

Results and Discussion

The count rate (expressed as counts per second, kcps) is in principle determined by the number of particles in the scattering volume, which has to exceed 100 (Wiener, 1991). This is equivalent to a count rate higher than 10 kcps for the present PCS equipment. From the laboratory experiments, it was found that the count rate was proportional to the colloidal concentration in the range 0.03–2, 0.1–2 and 0.1–7 mg/l, for the $\propto$–Fe_2O_3, Υ–$Al(OH)_3$ and SiO_2 reference colloids, respectively (Fig. 2).

Fig. 2 Relationship between the PCS count rate and the concentration of reference colloids. The initial size distributions were in the range of 50–270 and 10–75 nm for SiO_2 and Fe_2O_3, respectively, at pH 6.0 ±0.5□ , Fe_2O_3; ■, SiO_2. (a) Concentration range 0–20 mg/l, (b) concentration range 0–1.1 mg/l.

Conclusions

The following conclusions can be drawn:

The PCS technique can be adapted for characterization *in situ* of the colloidal fraction in natural waters. e.g., for concentration levels down to at least 0.1 mg/l.

This study clearly illustrates the importance of careful handling and preparation of a water sample in order to prevent any changes to its ....

Figure 5-2 Example from a laboratory report with background, research method, data, and conclusions.[2]

Introduction. In the introduction, give an idea of the overall topic and purpose of the report, and provide an overview of its contents. Remember: Avoid diving into the thick of the discussion; orient readers to the report first.

Background. Provide a discussion of the background leading up to the project. Typically, this involves discussing a research question or conflicting theories in

[2] Excerpts on the plastic-packaging experiment were drawn from "Taste and Odor Development in Water in Polyethylene Containers Exposed to Direct Sunlight" by Lucia Calvosa et al., *Water Research* (July 1994), pp. 1595–1600. Excerpts from the study of colloidal matter in groundwater were drawn from "Measurements *In Situ* of Concentration and Size Distribution of Colloidal Matter in Deep Groundwaters by Photon Correlation Spectroscopy," by Anna Ledin et al., *Water Research* (July 1994), pp. 1539–1545.

the research literature. Or, for example, you may want to apply an interesting discovery from another field to something in your own. Explore this background to give readers an idea why you are doing this work and to provide a context for understanding your own project. When you do, provide citations for the sources of information you use, using the standard bibliographic format (see Ch. 6).

Literature Review. Often included in the lab report is a discussion of the research literature related to your project. You summarize the findings of other researchers that have a bearing on your work. Again, use the standard bibliographic format.

Depending on the length and complexity of the report, all three of the elements just discussed—introduction, background, and literature review—may easily combine into one introductory paragraph without subheadings. But, regardless of their length, these three elements should occur at the beginning of a laboratory report, even if each is only a sentence or two.

Theory, Method, Procedure, Equipment. One of the next major sections in the laboratory report gives readers an idea of your theory or approach in relation to your project. For example, as a software engineer, you may suspect that computer users would prefer online documentation to printed documentation. To test this idea, you set up several computers in a laboratory and have a typical cross-section of computer users perform procedures you design. First, you'd want to discuss the common thinking on this subject—that computer users prefer printed material. Then you'd want to explain the method and procedures you use as well as the equipment and facilities. As you can see, this part of the report in particular enables readers to replicate your project.

Observations, Data, Findings, Results. In a laboratory project, you normally collect data and then organize and present it in a section of its own. The common approach is to present the data, often formatted into tables, graphs, or charts, without interpretive discussion.

Conclusions. In the conclusion section of a laboratory report, you draw conclusions based on the data you've gathered and explain why you think those conclusions are valid.

Implications and Further Research. A common section in laboratory reports involves exploring the implications of the data and conclusions, considering how they can be applied, and outlining further research possibilities.

As with the opening sections, these three sections—findings, conclusions, and implications—can be rolled into one. This does occur in shorter laboratory reports. In any case, the first two elements—the data and the conclusions—must be there.

Information Sources. At the end of most laboratory reports is a section that lists information sources used in the project. For entries in that list, use the bibliographic format.

FORMAT OF LABORATORY REPORTS

The laboratory report can be presented in memorandum format if it is short and addressed internally within an organization. However, it can be presented in the style of a formal report, with covers, table of contents, and appendixes. For reports over three or four pages, consider using the formal-report format.

SPECIFICATIONS

Specifications are descriptions of products or product requirements, or more broadly, they provide details for the design, manufacture, testing, installation, and use of a product. You typically see specifications in the documentation that comes with certain kinds of products, for example, CD players or computers. These describe the key technical characteristics of the item. But specifications are also written as a way of "specifying" construction and operational details of something. They are then used by people who actually construct it or go out and attempt to purchase it. When you write specifications, accuracy, precision of detail, and clarity are critical. Poorly written specifications can cause a range of problems and lead to lawsuits.

For these reasons then, specifications have a particular style, format, and organization. Here are some general recommendations:

- Find out what the specific requirements are for format, style, contents, and organization. If they are not documented, collect a big pile of specifications written by or for your company, and study them for characteristics like those in the following.
- Use two-column lists or tables (as shown in Figure 5-3) to list specific details. If the purpose is to indicate details such as dimensions, materials, weight, tolerances, and frequencies, regular paragraph-style writing might be less effective.
- For sentence-style presentation, use an outline style similar to the one shown in Figure 5-3. Make sure that each specification receives its own number–letter designation.
- In sentence-style specifications, make sure each specific requirement has its own separate sentence.

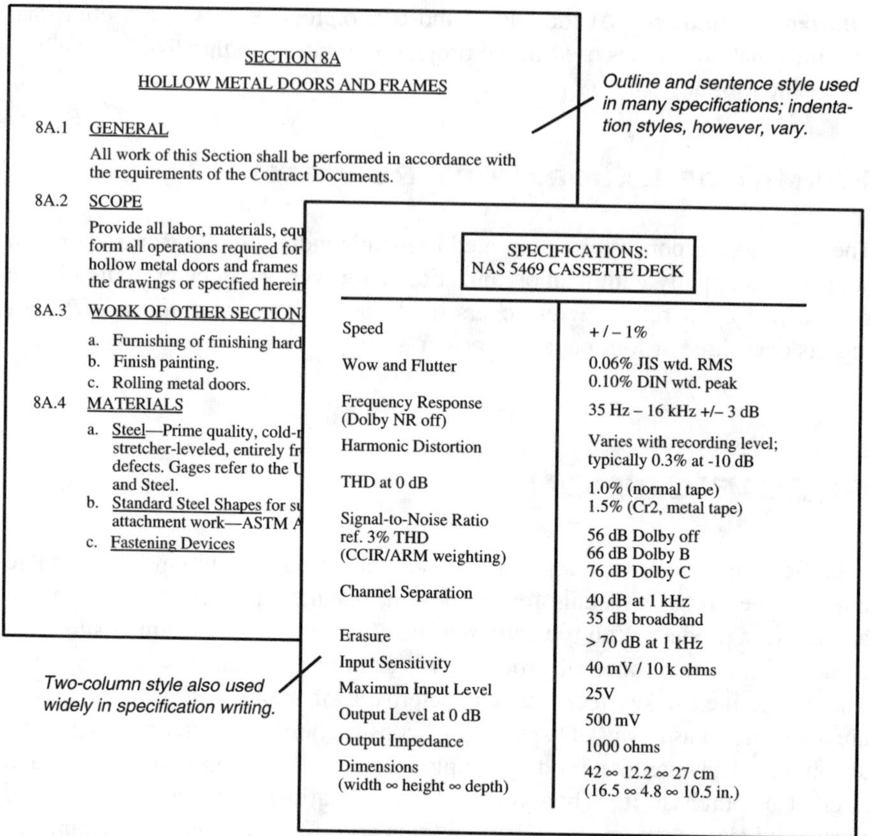

Outline and sentence style used in many specifications; indentation styles, however, vary.
Two-column style also used widely in specification writing.

SECTION 8A
HOLLOW METAL DOORS AND FRAMES

8A.1 GENERAL

All work of this Section shall be performed in accordance with the requirements of the Contract Documents.

8A.2 SCOPE

Provide all labor, materials, equ
form all operations required for
hollow metal doors and frames
the drawings or specified herei

8A.3 WORK OF OTHER SECTION

a. Furnishing of finishing hard
b. Finish painting.
c. Rolling metal doors.

8A.4 MATERIALS

a. Steel—Prime quality, cold-r
stretcher-leveled, entirely fr
defects. Gages refer to the U
and Steel.
b. Standard Steel Shapes for s
attachment work—ASTM A
c. Fastening Devices

SPECIFICATIONS:
NAS 5469 CASSETTE DECK

Speed	+ / – 1%
Wow and Flutter	0.06% JIS wtd. RMS 0.10% DIN wtd. peak
Frequency Response (Dolby NR off)	35 Hz – 16 kHz +/– 3 dB
Harmonic Distortion	Varies with recording level; typically 0.3% at -10 dB
THD at 0 dB	1.0% (normal tape) 1.5% (Cr2, metal tape)
Signal-to-Noise Ratio ref. 3% THD (CCIR/ARM weighting)	56 dB Dolby off 66 dB Dolby B 76 dB Dolby C
Channel Separation	40 dB at 1 kHz 35 dB broadband
Erasure	> 70 dB at 1 kHz
Input Sensitivity	40 mV / 10 k ohms
Maximum Input Level	25V
Output Level at 0 dB	500 mV
Output Impedance	1000 ohms
Dimensions (width ∞ height ∞ depth)	42 ∞ 12.2 ∞ 27 cm (16.5 ∞ 4.8 ∞ 10.5 in.)

Figure 5-3 Outline and two-column style used to present information in specifications. Graphics, tables, and lists are heavily used, but some details can only be provided through sentences and paragraphs.

- Use the decimal numbering system for each individual specification. This facilitates cross-referencing.
- Use either the open performance style or the closed restrictive style, depending on the requirements of the job. In the open or performance style, you can specify what the product or component should do, that is, its performance capabilities. In the closed style, you specify exactly what it should be or consist of.
- Cross-reference existing specifications whenever possible. Various government agencies as well as trade and professional associations publish specifications standards. You can refer to these standards rather than include the actual specifications details.

- Use specific, concrete language that identifies as precisely as possible what the product or component should be or do. Avoid ambiguity—that is, using words that can be interpreted in more than one way. Use technical jargon the way it is used in the trade or profession.
- For specifications to be used in design, manufacturing, construction, or procurement, use "shall" to indicate requirements. In specifications writing, "shall" is understood as indicating a requirement. (See the outline-style specifications in Figure 5-3 for examples of this style of writing.)
- Provide numerical specifications in both words and symbols: for example, "the distance between the two components shall be three centimeters (3 cm)."
- Writing style in specifications can be very terse: incomplete sentences are acceptable as well as the omission of obvious function words such as articles and conjunctions.
- Exercise caution with pronouns and relational or qualifying phrases. Make sure there is no doubt about the reference of words such as "it," "they," "which," and "that." Watch out for sentences containing a list of two or more items followed by some descriptive phrase—does the descriptive phrase refer to all the list items or just one? In cases like these, you may have to take a wordier approach for the sake of clarity.
- Use words and phrases that have become standard in similar specifications over the years. Past usage has proven them reliable. Avoid words and phrases that are known not to hold up in lawsuits.
- Make sure your specifications are complete—put yourself in the place of those who need your specifications; make sure you cover everything they will need.

Test your specifications by putting yourself in the role of a bumbling contractor—or even an unscrupulous one. What are the ways a careless or incompetent individual could misread your specifications? Could someone willfully misread your specifications in order to cut cost, time, and quality? Obviously, no set of specifications can ultimately be "foolproof" or "shark-proof," but you must try to make them as clear and unambiguous as possible.

Contents and Organization of Specifications

Organization is critical in specifications—readers need to be able to find one or a collection of specific details. To make individual specifications easy to find, use headings, lists, tables, and identifying numbers as discussed previously. A certain organization of the actual contents to enable this retrieval is also standard:

- *General description:* Describe the product, component, or program first in general terms—administrative details about its cost, start and completion dates, overall description of the project, scope of the specifications (what you are not covering), anything that is of a general nature and does not fit in the part-by-part descriptions.
- *Part-by-part description:* In the main text, present specifications part by part, element by element, trade by trade—whatever is the logical, natural, or conventional way of doing it.
- *General-to-specific order:* Wherever applicable, arrange specifications from general to specific.

GRAPHICS IN SPECIFICATIONS

In specifications, use graphics wherever they enable you to convey information more effectively. For example, in the specifications for a cleanroom for production of integrated circuits, drawings, diagrams, and schematics convey some of the information much more succinctly and effectively than sentences and paragraphs. See the example of a graphic used in specifications writing in Figure 5-4. As you prepare to write a set of specifications, look ahead as best you can to determine which formats—graphics, tables (or lists), and sentences—will be the most succinct, exact, and practical for each of the sections of your specifications.

PROGRESS REPORTS

Another common type of report is variously called the progress report, status report, interim report, quarterly report, monthly report, and so on. Its job is to present to your clients the status of the work you are doing for them. You supply these details to enable your clients to act as manager or executive of the project—to enable them to modify it or even cancel it if the need arises. In this situation, you are the *supplier* of the work of the project; the recipient of the work is the *customer*—the individual or organization that requires the work. Your client may be internal to your organization, for example, a work supervisor; or external, for example, a customer with whom you have a contract.

To understand the function of progress reports, it's helpful first to understand the projects for which they are written. In a project of any size, length, or

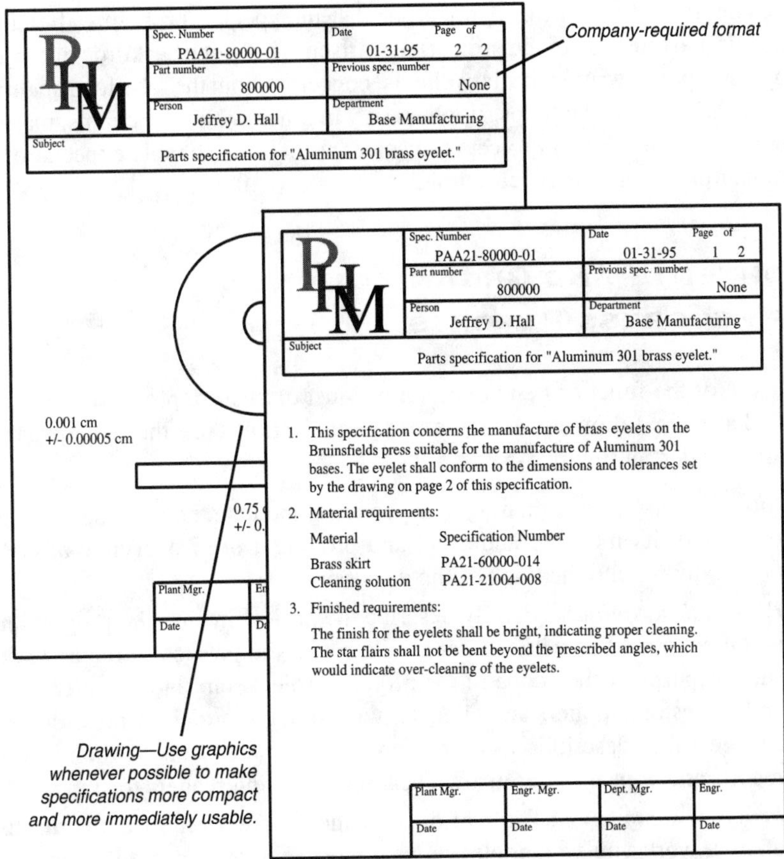

Figure 5-4 Graphics and tables used for the presentation of information in specifications.[3]

importance, there are bound to be changes, new and additional requirements, problems, and miscommunications. In highly competitive new-product development, these potentials are greatly magnified. Large development organizations of five hundred to a thousand people must work fast and be able to change quickly in order to deliver a product that is attractive to the marketplace. As a supplier, you may be a critical part of this process.

In this environment, clients may worry that the work is not being done properly, on schedule, or within budget. Suppliers, on the other hand, may worry that clients will not like how the project is developing, that new requirements jeopardize the

[3] Example specifications drawn from work done by Jeffrey D. Hall, engineering student, University of Texas at Austin, 3 January, 1989.

schedule and budget for the project, or that unexpected problems affecting the schedule and outcome of the project have arisen. This is the backdrop on which the progress report is written: it allays clients' concerns about the schedule, quality, and cost of their projects; it helps suppliers stay in touch with their clients, maintain a professional image, and protect themselves from unreasonable expectations and mistaken or unwarranted accusations.

CONTENTS AND ORGANIZATION OF PROGRESS REPORTS

Because of the functions and expectations just discussed, progress reports typically have the following contents and organization (see the excerpts from a progress report in Figure 5-5):

- *Introduction:* As with any report, you first indicate the purpose and topic of the report, its intended audience, and provide a brief overview of contents *before* diving into the thick of the discussion.
- *Project description:* Briefly describe the project in case the progress report is routed to readers who are not familiar with the project. Summarize details such as purpose and scope of the project, project start and completion dates, and names of suppliers and clients involved in the project. Unless the project changes, this description can become "boilerplate" text in future progress reports and appear under its own heading, enabling readers to skip it.
- *Progress summary:* The real meat of the progress report is the discussion of what work you've completed, what work is in progress, and what's yet to come. This discussion can be handled several ways:
  - *Time-periods approach:* You can summarize work completed in the previous period, work under way in the current period, and work planned for future periods.
  - *Project-tasks approach:* You can summarize which tasks in the project have been completed, which tasks are currently under way, and which tasks are planned for future work.
  - *Combined approach:* You can combine these approaches by dividing the section on previous-period work into summaries of the work you did on individual tasks. Or, you can divide project-task sections into summaries of work completed, in progress, or planned for each task.

Which of these approaches you use is strictly dependent on the nature of your project and the requirements of your client. For simpler projects, however,

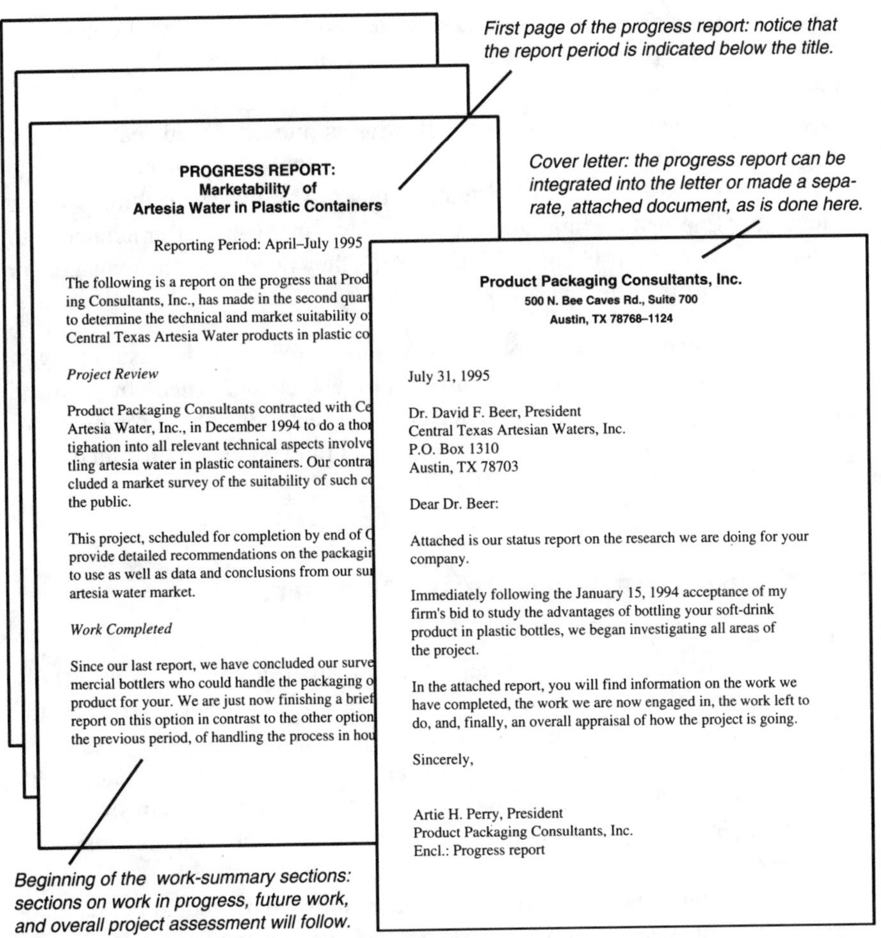

First page of the progress report: notice that the report period is indicated below the title.
Cover letter: the progress report can be integrated into the letter or made a separate, attached document, as is done here.
PROGRESS REPORT:
Marketability of
Artesia Water in Plastic Containers
Reporting Period: April–July 1995
The following is a report on the progress that Prod
ing Consultants, Inc., has made in the second quar
to determine the technical and market suitability o
Central Texas Artesia Water products in plastic co
Project Review
Product Packaging Consultants contracted with Ce
Artesia Water, Inc., in December 1994 to do a thor
tighation into all relevant technical aspects involve
tling artesia water in plastic containers. Our contra
cluded a market survey of the suitability of such co
the public.
This project, scheduled for completion by end of C
provide detailed recommendations on the packagi
to use as well as data and conclusions from our su
artesia water market.
Work Completed
Since our last report, we have concluded our surve
mercial bottlers who could handle the packaging o
product for your. We are just now finishing a brief
report on this option in contrast to the other option
the previous period, of handling the process in hou
Product Packaging Consultants, Inc.
500 N. Bee Caves Rd., Suite 700
Austin, TX 78768–1124
July 31, 1995
Dr. David F. Beer, President
Central Texas Artesian Waters, Inc.
P.O. Box 1310
Austin, TX 78703
Dear Dr. Beer:
Attached is our status report on the research we are doing for your company.
Immediately following the January 15, 1994 acceptance of my firm's bid to study the advantages of bottling your soft-drink product in plastic bottles, we began investigating all areas of the project.
In the attached report, you will find information on the work we have completed, the work we are now engaged in, the work left to do, and, finally, an overall appraisal of how the project is going.
Sincerely,
Artie H. Perry, President
Product Packaging Consultants, Inc.
Encl.: Progress report
Beginning of the work-summary sections: sections on work in progress, future work, and overall project assessment will follow.

Figure 5-5 Cover letter and first page of a progress report. If your progress report is short, you can incorporate the report into the letter, making it one continuous document (for an example of this approach, see the letter and memo proposals in Figure 5-6).

the time-periods approach may work best. The project-tasks approach works well when the project has a number of semi-independent tasks on which you are working more or less concurrently.

- *Problems encountered:* In this section, you go on record about the problems that have arisen in the project, problems you think may jeopardize the quality, cost, or schedule of the project.

- *Changes in requirements:* In this section, you keep a history of changes in the project as you understand them: for example, schedule shifts, new requirements, and so on.
- *Overall assessment of the project:* In what is often the final section of the progress report, you give a general opinion as to how the project is going. In this section, resist the opposing temptations to say that everything's going along just fine or to whine about every minor annoyance. Remember your job is to provide your clients with the details they need to act as managers or executives of the project as a whole.

Of course, other sections may also be required: for example, a summary of financial data on the project or the results of product testing. When you plan and write progress reports, be alert to the needs and expectations of your audience—in this case, those customers or supervisors on whom you depend for income or employment.

FREQUENCY OF PROGRESS REPORTS

If you're not sure whether progress reports are required for a project, especially short projects, check with your supervisor or client. Remember that the "progress report" may be nothing more than a quick e-mail note briefly describing where you are on the project. It's a healthy impulse to avoid unnecessary work, but keep in mind that progress reports, when appropriate, can strengthen your professional image. They keep you closer to your client, and they may help eliminate some unfortunate surprises.

The schedule for progress reports may be established by your supervisors or clients. If it's not, your sense of the project and the requirements of the client should dictate how many progress reports there should be and how often they should be delivered. Typically, progress reports are sent at the end of every month or every quarter. The larger the project, the more formally defined these requirements will be and the more formal the progress report will be.

FORMAT FOR PROGRESS REPORTS

For large projects, progress reports can be lengthy, 100-page, bound, formal reports. Even so, the contents and organization are essentially the same as previously discussed. The formal elements include title page, table of contents, abstracts, appendixes, and so on. It's more likely, however, that your progress report can fit easily into a business letter or memorandum.

PROPOSALS

One of your most important tools as an engineer, particularly as a *consulting* engineer, is the proposal. With it, you get work, either for yourself, if you're an independent, or for the company that employs you.

If you explore the literature on proposals, you'll see that they are defined in many different ways. In this book, however, the proposal is something quite specific: it is a bid, offer, or request to do a project plus any supporting information necessary to gain approval or acceptance to do the project. Proposals sometimes must convince the recipient that the project needs to be done, but proposals must *always* convince the recipient that the proposer is the right individual or organization to do the project.

In the typical proposal scenario, an organization[4] sends out a request for proposals (RFP) to do a certain project. These RFPs can be sent out in various ways: by publication in newspapers, professional journals, or specialized periodicals such as the *Commerce Business Daily;* by mail to a select list of vendor organizations; or by various informal means such as telephone or e-mail. Organizations interested in doing the project submit proposals in which they present their qualifications and make a case for themselves as a good choice. The recipient of the proposals selects one of the proposals and enters into contract negotiations. Once that is accomplished, the organization that won the project can get down to work.

As you can see, proposal writing is a competitive affair. You must highlight your organization's strengths; you must make a good case for your company as the right one for the project.

TYPES OF PROPOSALS

Proposals are commonly divided into two types based on whether the recipient requested them:

- *Solicited.* If an organization issues a request for proposals, the proposals are said to be "solicited"—they have been requested.
- *Unsolicited.* Individuals and companies often initiate proposals without formal request from the recipients. They may see that an individual or organization has a problem or opportunity. When the proposal is unsolicited, you,

[4] *Organization*, as the term is used here, refers to for-profit companies, not-for-profit organizations, and government agencies. All of these organizations, both commercial and noncommercial, request and submit proposals.

the proposal writer, have to do the additional work of convincing the recipient that the project needs to be done.

Proposals can also be divided according to the context in which they occur:

- *Internal.* If you address your proposal to someone within your organization, the format and contents may change significantly. The memo format is usually appropriate, and sections like qualifications and costs may not be necessary.
- *External.* If you address your proposal to some other individual or organization outside of your own, you must use some combination of the business-letter and formal-report formats.

ORGANIZATION AND CONTENT OF PROPOSALS

The typical sections in a proposal are as follows (see the proposal excerpts in Figure 5-6 in which some of these sections are illustrated):

- *Introduction.* In the first paragraph or section of a proposal, make reference to some prior contact with the recipient of the proposal or your source of information about the project. Identify the information that follows as a proposal (in other words, state the purpose). Also, give a brief overview of the contents of the proposal.
- *Background.* In most unsolicited proposals, you'll want to discuss the problem or opportunity that caused you to write the proposal. In solicited proposals, this may not be necessary: The party requesting proposals probably knows the problem very well. Still, a background section even in a solicited proposal can be useful in demonstrating that you fully understand the situation or in exploring your interpretation of it.
- *Actual proposal statement.* In most proposals, you include a short section in which you state explicitly what you are proposing to do. Proposals often refer to many possibilities, which can create some vagueness about what's actually being offered. You may also need a scope statement—an explicit statement about what you are *not* offering to do.
- *Description of the work product.* Many proposals need a section in which the proposed project—in other words, the results of the work—is described. This might be a constructed building, a program design, blueprints or plans, or even a 40-page report. The point is to provide details on what the recipient is getting.
- *Benefits and feasibility of the project.* As a way of promoting the project to the recipient, some proposals discuss the likely benefits of the project. Others discuss how likely those benefits are. This is particularly true in unsolicited

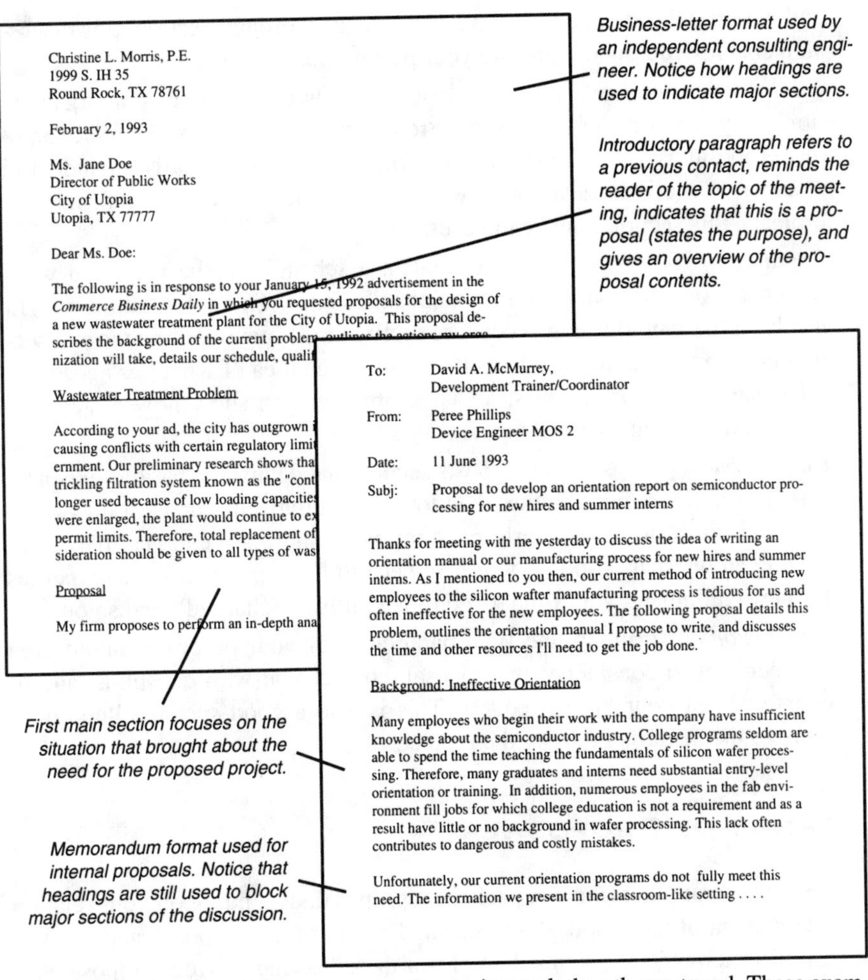

Christine L. Morris, P.E.
1999 S. IH 35
Round Rock, TX 78761
February 2, 1993
Ms. Jane Doe
Director of Public Works
City of Utopia
Utopia, TX 77777
Dear Ms. Doe:
The following is in response to your January 15, 1992 advertisement in the Commerce Business Daily in which you requested proposals for the design of a new wastewater treatment plant for the City of Utopia. This proposal describes the background of the current problem, outlines the actions my organization will take, details our schedule, quali
Wastewater Treatment Problem
According to your ad, the city has outgrown
causing conflicts with certain regulatory limi
ernment. Our preliminary research shows tha
trickling filtration system known as the "cont
longer used because of low loading capacitie
were enlarged, the plant would continue to e
permit limits. Therefore, total replacement of
sideration should be given to all types of was
Proposal
My firm proposes to perform an in-depth ana
To: David A. McMurrey, Development Trainer/Coordinator
From: Peree Phillips, Device Engineer MOS 2
Date: 11 June 1993
Subj: Proposal to develop an orientation report on semiconductor processing for new hires and summer interns
Thanks for meeting with me yesterday to discuss the idea of writing an orientation manual or our manufacturing process for new hires and summer interns. As I mentioned to you then, our current method of introducing new employees to the silicon wafter manufacturing process is tedious for us and often ineffective for the new employees. The following proposal details this problem, outlines the orientation manual I propose to write, and discusses the time and other resources I'll need to get the job done.
Background: Ineffective Orientation
Many employees who begin their work with the company have insufficient knowledge about the semiconductor industry. College programs seldom are able to spend the time teaching the fundamentals of silicon wafer processing. Therefore, many graduates and interns need substantial entry-level orientation or training. In addition, numerous employees in the fab environment fill jobs for which college education is not a requirement and as a result have little or no background in wafer processing. This lack often contributes to dangerous and costly mistakes.
Unfortunately, our current orientation programs do not fully meet this need. The information we present in the classroom-like setting . . . .
Business-letter format used by an independent consulting engineer. Notice how headings are used to indicate major sections.
Introductory paragraph refers to a previous contact, reminds the reader of the topic of the meeting, indicates that this is a proposal (states the purpose), and gives an overview of the proposal contents.
First main section focuses on the situation that brought about the need for the proposed project.
Memorandum format used for internal proposals. Notice that headings are still used to block major sections of the discussion.

Figure 5-6 Excerpts from two proposals, one internal, the other external. These examples integrate the cover letter (or memo) and the proposal proper into one continuous document.[5]

proposals where the recipient must be convinced that the project is necessary in the first place.

- *Method or approach.* In some of your proposals, you'll need a section that explains how you plan to go about the project, even the theory behind your approach. For some projects, people need to know how the work will be done

[5] Example proposals drawn from work by Peree Phillips and Christine Morris, students at Austin Community College, 22 January, 1992.

and why it will be done that way. As in the background section, this discussion enables you to demonstrate your professional expertise.

- *Qualifications and references.* Most proposals list the proposing organization's key qualifications, along with references to past work. This section of the proposal is like a miniresume. Large proposals actually include full resumes of the individuals who will work on the project. In small internal projects, this section may be unnecessary.
- *Schedule.* The proposal should contain a schedule of the projected work with dates or a time line for the major milestones. This information may fit nicely in the methods and procedure section, or it may work better in a section of its own. Again, this gives the recipient an idea of what lies ahead and a chance to ask for changes; and it enables you to show how systematic, organized, and professional you are.
- *Costs.* Most proposals have a costs section that details the various expenses involved in the project. Rather than toss out a lump sum, break it down into different kinds of labor, hourly rates for each, and other charges. If you are writing an internal proposal, you may need to list supplies needed, expenses for new equipment, your time (even though it is not charged), and so on.
- *Conclusion.* Normally, the final paragraphs of your proposal should urge the recipient to consider your proposal, contact you with questions, and of course accept your bid or request. This is also a good spot to allude once more to the benefits of doing the project.

FORMAT OF PROPOSALS

There are several ways you can package a proposal, depending on your relation to the recipient of the proposal, the size and nature of the proposal, and the way it will be used by the recipients. Use one of the following for your proposal:

- *Memorandum format.* If your proposal is short (under four pages), and if it's addressed to someone within your company, use a simple memo format (see Figure 5-6). Include headings as you normally would for any other document.
- *Business-letter format.* If your proposal is short but is addressed to someone *outside* your organization, use a business letter (also illustrated in Figure 5-6). Again, include headings as you normally would.
- *Separate proposal with cover memo.* If your proposal is long (over four pages), if it's addressed to people *within* your own company, and if you envision it being passed around among various reviewers, make it a separate document with its own title. Attach a cover memo to the front of it; in the memo, restate the key elements of the introduction and the conclusion. (See Figure 5-7, but picture the letter reformatted as a memorandum.)

- *Separate proposal with cover letter.* If your proposal is long, if it's addressed to people *outside* your company, and if you envision it being passed around among various reviewers, make it a separate document with its own title. Attach a cover letter to the front of it; in the letter, restate the key elements of the introduction and the conclusion. (See Figure 5-7.)

INSTRUCTIONS

You may often find yourself having to write step-by-step procedures for employees, colleagues, customers, or clients. In such instructions, you explain how to assemble, operate, or troubleshoot some new product your team is working on. Or you may need to explain in writing how to operate equipment around the office, laboratory, or site, or how to perform other kinds of procedures.

SOME PRELIMINARIES

We've all had upsetting experiences with poorly written instructions. It's surprising that experienced, reasonably intelligent, well-intentioned people can write terrible instructions. The following is not a foolproof guide for writing instructions, but it does show how instructions are commonly organized and presented. Critical in instructions writing is putting yourself in your readers' place, making no unwarranted assumptions about their background or knowledge, and providing them everything they need to successfully complete the procedure.

Understand that there is an important difference between instructions and product specifications. In rushed development cycles, product specifications are sometimes used, with little revision, as instructions. That's unfortunate because specifications (discussed previously in this chapter) normally do not function well as instructions. Specifications approach the product as a group of features and functions—not in terms of tasks. For example, a product specification for a phone-answering machine might discuss the function of each button and the capabilities of the machine, but those descriptions may not enable the ordinary user to know which buttons to push to record a new greeting.

Critical in preparing to write instructions is audience analysis—identifying the relevant characteristics of that group of readers most likely to use your instructions. (For a full discussion of this task, see Ch. 2.) With most instructions, you must decide what you expect your readers to know and what you will explain in your instructions. For example, in explaining how to install a computer program, you have to decide whether to assume readers understand some basics about diskettes, directories, or files.

PROPOSAL
to
Develop a Guide for Writing
Policies and Procedures

The following is a proposal t
business people in writing po
Included in this proposal is a
guide like this, a description
schedule for its development

Problem: Businesses without

Many small businesses in Au
handbook or guide for emplo
dures governing that business
usually ill-equipped to devel
are the pressure of their norm
ty with this type of documen
writing experience.

However, the lack of such op
various kinds of problems fo
productivity, loss of sales op
lawsuits by unhappy employ

Proposal

Morris Business Consultants
book for distribution to Aust
of Commerce. This handboo
local business peopler in dev
procedures manuals.

MORRIS BUSINESS CONSULTANTS
1005 Twin Towers Suite 105
Austin, TX 78761

May 25, 1993

Mr. Patrick H. McMurrey
Chamber of Commerce
100 West 1st Street
Utopia, TX 78777

Dear Mr. McMurrey:

I enjoyed meeting with you this past Monday, May 21st, concerning your interest in a manual on how to write policies and procedures. Local-area businesses do need help in this very important area of their operations. With the lack of good practical guides available commercially, developing and making one available through the Chamber will serve a great need.

The attached proposal outlines the details of our conversation concerning the content of the guide, costs, schedules, and so forth. I've added my qualifications for review by your other colleagues.

If you have any questions, please feel free to call me at 455-1122 during business hours.

Respectfully,

Gayle Morris

Gayle Morris,
Business Consultant

Attachment: Proposal

Figure 5-7 Excerpts from a proposal that uses a cover letter. The proposal proper uses a title at the top of the page and repeats some of the contents of the cover letter (in case the letter is separated from the proposal).[6]

COMMON SECTIONS IN INSTRUCTIONS

Introduction. In the introduction to most instructions, include some combination of the following:

- *Subject or focus of the instructions:* Indicate which specific procedures or tasks you'll actually explain.
- *Product:* If you are providing instructions for a product, give some brief description or overview of it.

[6] Example proposal drawn from work done by Gayle Morris, student at Austin Community Col-[illegible], 22 January, 1992.

- *Audience:* Indicate the knowledge or background, if any, that readers need in order to understand your instructions. If no special background is needed, indicate that as well.
- *Overview:* Briefly list the main contents of the instructions; for example, list the major tasks or procedures to be presented.

Special notices. In many instructions, you'll see specially formatted notices for warnings, cautions, and dangers. Often these will appear right in the introduction as well as in the body of the instructions at those points where individual special notices are relevant. Lack of these special notices in instructions can lead to lawsuits when readers injure themselves or lose money.

Style and format of special notices vary widely, but here's a recommended approach.

Note: Used for emphasizing special points that might otherwise be overlooked.

Warning: Used for alerting readers to a potential for ruining the outcome of the procedure or damaging the equipment.

Caution: Used for alerting readers to the possibility of minor injury.

Danger: Used to alert readers to the possibility of serious or even fatal injury.

See Figure 5-8 for examples of these special notices. Notice that the serious ones are placed *before* the point at which readers might wreck their procedure, ruin their equipment, hurt somebody, or blow themselves up!

Background. Some instructions benefit greatly by an introductory discussion of concepts or theory related to the procedures. For certain complex tasks, it helps readers to know conceptually what they are doing and why they are doing it. Background can enable them to figure out other procedures for themselves without resorting to instructions at all. But make sure your background is closely related to the procedural discussion. A common problem with instructions is that the background often has no immediate relation to the tasks it is supposed to elucidate. To prevent this problem, write background you think you need only *after* you've written the procedures.

Equipment and supplies. In most instructions, you'll need to list the supplies and equipment that readers must gather before they begin. *Supplies* are the consumable items used in the process such as paper, flour, glue, sandpaper, eggs, milk, nails, paint, paint thinner, and sugar. *Equipment* is the tools and machinery that are needed, such as screwdrivers, hammers, mixing bowls, cake pans, and so on. For some instructions, it's not enough merely to list equipment and

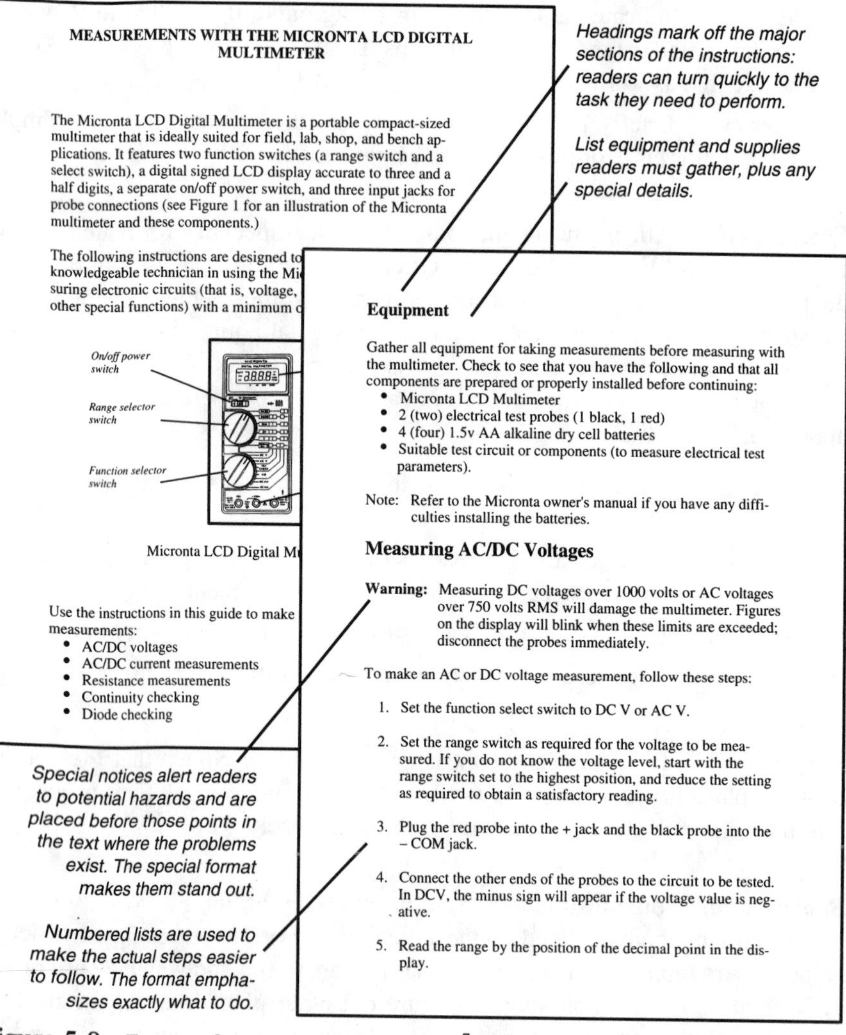

Figure 5-8 Excerpts from a set of instructions.[7] Headings, lists, graphics, and special notices are critical elements of good instructions—but so is good, clear writing.

supplies. You may also need to list specific sizes, brands, types, models, and so on.

In some instructions, you may need to orient readers to the equipment they will be using. For instructions on a top-of-the-line photocopier or fax machine, you'd need to familiarize readers with the buttons and other sorts of controls on the machine. (But don't make the mistake of assuming that these *are* the instruc-

[7] Example instructions drawn from work done by Robert Hutchinson, student at Austin Community College, 27 June, 1991.

tions. There may be a LISTEN button on the telephone-answering machine, but what are the steps for listening to a message?)

Structure of the instructions. Before you dive into the step-by-step discussion of the procedure, identify the tasks involved in that procedure. Your instructions may have only one task—a simple series of steps that the reader must perform in sequence. However, your instructions may involve several tasks that can be performed in practically any combination or order. For example, changing the oil in a car involves one task, a relatively simple series of steps that must be performed in order—otherwise, you'll have oil all over the driveway or a burned-up engine! Operating a phone-answering machine involves numerous tasks, some of which you perform only once, others every day—for example, recording your greeting, playing back messages, saving or deleting messages, forwarding messages, and using an alternate greeting.

Discussion of the steps. As for the step-by-step discussion of how to perform the tasks, be aware of some issues involving writing style, format, headings, and content:

- *Imperative writing style:* Notice that many sentences in instructions use the imperative (for example, "Press the Enter key" or "Calculate the square footage in the lot") or are phrased with the word "you" (for example, "You should check the temperature of the . . ."). Use this "in-your-face" style of writing—address readers directly, get their full attention, and be as clear and straightforward as possible about what they are supposed to do.
- *Supplemental explanation:* For individual steps in instructions, additional explanation may be needed—in many cases, you must explain to readers why they should or should not do something. You may also need to define potentially unfamiliar terms or describe how things look before, during, or after individual steps. But notice also that instruction writers often keep specific instructions separate from the supplemental explanation, perhaps by skipping a line or using a different type style.
- *Special format:* In the discussion of steps, vertical numbered lists are often used. Obviously, this helps readers follow the procedure and clearly defines each specific action they must perform.
- *Headings:* Notice finally that well-written instructions use headings. Headings help readers find the list of equipment and supplies, the background information, the troubleshooting tips. Headings help readers find the different tasks that make up the procedure. Instructions for a phone-answering machine would have headings for the sections on recording a greeting, playing back messages, saving or deleting messages, forwarding messages, and using an alternate greeting—these help readers locate the information they need quickly.

USE OF GRAPHICS IN INSTRUCTIONS

Graphics are often critical in instructions. Sometimes words simply cannot convey details about key objects and key actions in the procedure. You've probably had experiences with instructions in which graphics were painfully inadequate: just how does part A fit into part B? In fact, graphics have generally the same problems that text does: They can be inaccurate—an illustration of a part may not correspond with what came in the package; they can be incomplete—key details may be missing in the illustration; and they can be unclear—the quality or orientation of the drawing can be so poor that you can't discern what is what.

RECOMMENDATION REPORTS

A *recommendation report* evaluates or promotes an idea. For example, you could write a recommendation report (sometimes even called a proposal) that enthusiastically endorses telecommuting as an option for your fellow employees. Or your management might ask you to study the feasibility of telecommuting and to make recommendations whether to implement such a program. And, finally, you might be asked to compare various software products for use in the company telecommuting program and then to recommend one of them. The common element for these reports is a recommendation and, as the following discussion will show, a comparative discussion that supports those recommendations. Where you work, it may be called a recommendation report, evaluation report, feasibility report—or even a proposal. But the essential structure is the same for all—comparing options and recommending one.

SOME DISTINCTIONS

A recommendation report, as its name indicates, makes a recommendation about plans, products, or people. In its simplest form, it defines certain requirements (often called criteria), compares two or more options, and recommends one. Of course, there can be other elements. For example, you might be asked to investigate e-mail software and recommend one package for use in your company. Part of your job would be to survey the field of e-mail software and narrow the field to four or five strong candidates. In some recommendation reports, elements of the feasibility report enter in: Part of your job might be to study the technical, economical, and human practicality of an idea before focusing on which is the best.

Typically, the terms proposal, feasibility report, evaluation report, and recommendation report are used interchangeably. Don't expect much precision in real-world usage of these terms. To make the distinctions used in this book:

Recommendation report. Compares two or more options against each other (and against certain requirements) and then makes a recommendation.

Evaluation report. Compares an idea or thing against criteria or requirements as a means of determining its worth or value. In this type of report, there may be a recommendation, but the essential element is the statement on the worth or value of the idea or thing.

Feasibility report. Studies a project in terms of its economic, technical, or social practicality, and then recommends whether the project should be initiated.

Proposal. Makes a bid or seeks approval to do a project and then supplies supporting information on the bidder's qualifications. It may promote an idea, but that is secondary, at least in this book, to its primary task of landing a contract or getting approval. (See the section on proposals earlier in this chapter.)

As you can see, each of these types, except for the proposal, works toward some sort of endorsement, recommendation, or value judgment.

CONTENTS AND ORGANIZATION OF RECOMMENDATION REPORTS

To understand the contents and organization of this type of report, bear in mind that, while its job is indeed to provide a judgment or recommendation, it should also provide the data and conclusions so that readers can decide for themselves whether the recommendations are justified.

Introduction. As with any introduction to a report, indicate the purpose of the document. Indicate right up front that the point of this report is to recommend something for some specific use or situation. Indicate the audience—the intended readers of the report and what technical background, if any, they need (and if none is needed, say so). Also include an overview; provide a brief list of the contents of the report.

Background on the situation. In some recommendation reports, you may need to discuss the situation in which the recommendation report is needed. The immediate audience may know perfectly well what the situation is, but your report may get passed around to others who don't. This section also provides history that may prove a helpful memory jogger. And remember, if you use headings, readers who know this background can simply skip over it.

Requirements. In practically any situation where a recommendation is needed, there are requirements such as cost, operational features, size, flow rates, weight, and so on. Consider the example of selecting an e-mail software package for the company: What are the specifications? Ease of use? Versions for Macintosh and PC machines? File transfer capability? Bulletin board features? In your recommendation *study*, you'll probably nail down these requirements. In your recommendation *report*, you'll list and describe these requirements. Readers can then consider these requirements and decide for themselves whether they agree.

Technical background. For some recommendation reports, it may be necessary to dip very briefly into some technical discussion. If you were comparing CD-ROM products for office use, you might need to discuss 8- and 16-bit technology, triple or quadruple spin, sampling, and other related technical concepts. As with most technical background, it's usually best to write this section later. Write the heart of the recommendation report—the comparisons, conclusions, and recommendations—then go back and look for terms or concepts that need explanation.

Description. In some recommendation reports, you may need to describe the options you are comparing. In this section, the description is neutral, with no comparisons provided. For example, if you were making a recommendation on laser printers, you might want to describe each of the finalists separately in terms of its size, dimensions, operating features, warranties, upgrade possibilities, and so on. Again, as with the technical-background section, this section may not be necessary; however, if it is, it ought to be written only *after* the main sections.

Point-by-point comparisons. Comparisons constitute one of the three main sections of a recommendation report. In comparison sections, you focus the discussion on specific *points*, such as cost, ease of use, warranties, and features. For example, you might have a section on cost where you compare the cost of each of your choices. Usually, it's not a simple matter of one being the cheapest, and another being the most expensive. Things get blurred by special features and service plans that can be added on. In these cases, help readers by untangling the complexities for them and pointing to the best choice.

Remember, you are writing these comparisons so that readers can see your logic—how you reached your conclusions. You're giving readers a chance to disagree with your thinking and to come up with their own conclusions and recommendations.

For some comparisons, you may need to use a weighting factor so that the different requirements or categories receive more or less consideration according to their importance.

Conclusions. The conclusions section of a recommendation report is a summary of the conclusions you reached during the comparison section. Notice in

Background: Need for Laptops

Software Designs, Inc., owns numerous desktop and midrange computers which are adequate for the company's office needs. However, there are many occasions when smaller laptop- or notebook-style computers would be useful. Often, employees need to take a computer with them to customer sites—either to prospective customers' offices for demonstration of our company's past work or to current customers' offices for demonstrations of work in progress. Also, there are many occasions when it would be advantageous for employees to take a laptop or notebook computer home with them or on other kinds of business trips.

Requirements for the Laptops

Any decision to purchase laptop or noteb
must take into account the following crit

- Personal computer and DOS comp
- Size, weight, portability, and batte
- RAM memory size, hard disk spac
- Quality and readability of the displ capabilities
- Modem capabilities
- Cost of the machines.

The laptop computers will need to have
graphics capabilities are almost requirem
show off our software in the best possibl
fast machines—486s with at least 8MB
reliable: at least 2 hours of operation are
because we will be purchasing these com
have to be under $1000.

2

Background and requirements sections: in the background section, show what the needs are; in the requirements section, list the criteria you'll use in making your recommendations.

Memory. Working memory for the Shiva HB60/S is 4MB, expandable to 8MB. The Quantum N400 begins with 8MB of working memory and can be expanded to 12MB. The ICON S/6000 starts at 6MB and can only expand to 8MB. The ICON S/6000 should be sufficient for our needs, however.

Screen quality. The Shiva HB60/S has far and away the best screen quality of the three machines, and this probably accounts for its cost. You get vivid color and graphics; plus the display is easy to read. The Quantum N400 offers a fairly conventional display, with no color but an adequately readable screen. The ICON S/6000 is clearly the least acceptable: it lacks color and graphics capability, plus the screen can be hard to read in certain environments.

Cost. The list price of the Shiva HB60/S is $1095 but has been advertised at $1198. The Quantum N400 has a list price of $1400 but has been advertised at $999. The ICON S/6000 lists at $999 but has been advertised at $700. The ICON is obviously the cheapest machine, but it may not meet all of our requirements.

Table of Laptop Computer Rankings

	Shiva HB60/S		Quantum N400		ICON S/6000	
Compatibility	yes	3	yes	3	yes	3
Weight (lbs.)	11.11	2	14.0	1	6.2	3
Battery life (hrs.)	3	1	5.5	3	4.25	2
RAM memory (MB)	4	1	8	3	6	2
Hard drive (MB)	40	2	40	2	25	3
Screen quality	good	3	ave.	2	poor	1
Advertised price	$1198	1	$999	2	$700	3
		13		16		17

3

Comparative discussion: the options are compared category by category, for example, costs, warranties, special features. For each category, state a conclusion as to which is the best.

Summary table: summarize the comparative information in table format as well as providing it in textual discussion.

Figure 5-9 Key elements of a recommendation report: discussion of requirements (criteria), category-by-category comparative discussion, and summary tables.

the example in Figure 5-9 that the comparison of costs ends with a conclusion—which choice is least expensive. Notice that item 7 in the conclusion section shown in Figure 5-10 echoes this conclusion on costs.

In some recommendation reports, there may be no clear or obvious choice. One choice may be the cheapest; another may be the most reliable; another may be the easiest to use; still another may have far more functions and features. These are the *primary conclusions*. But how do you pick a "winner" when they conflict? If you've defined them carefully, your requirements should point the way to the final recommendation. Requirements enable you to state *secondary conclusions*—conclusions that resolve conflicting primary conclusions. Notice the secondary conclusions in Figure 5-10.

Recommendations. The recommendations section simply summarizes what has probably become obvious—which choice is recommended. The example in

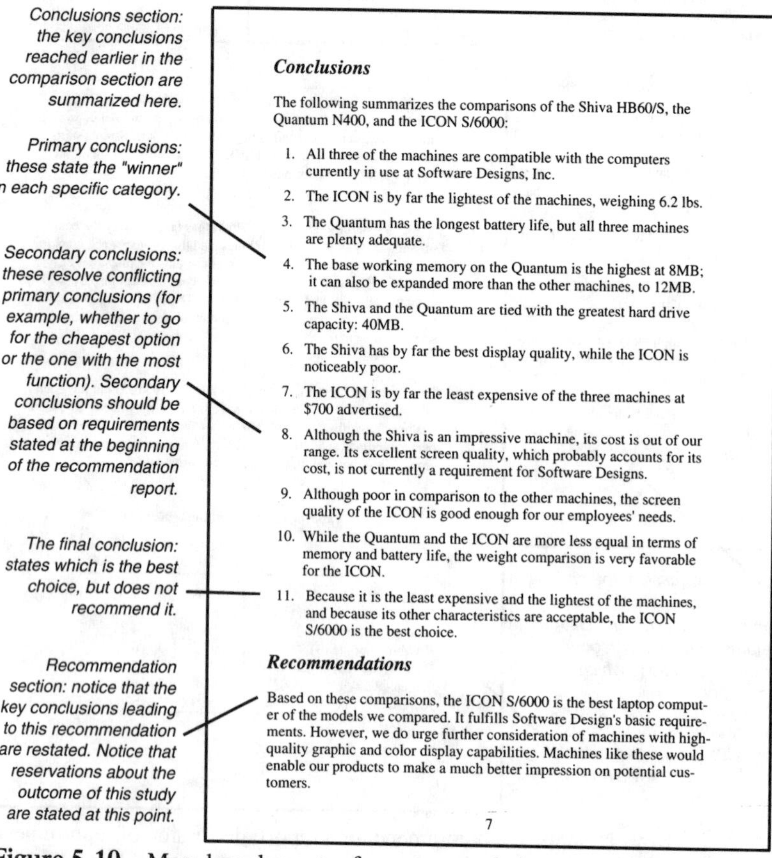

Conclusions

The following summarizes the comparisons of the Shiva HB60/S, the Quantum N400, and the ICON S/6000:

1. All three of the machines are compatible with the computers currently in use at Software Designs, Inc.
2. The ICON is by far the lightest of the machines, weighing 6.2 lbs.
3. The Quantum has the longest battery life, but all three machines are plenty adequate.
4. The base working memory on the Quantum is the highest at 8MB; it can also be expanded more than the other machines, to 12MB.
5. The Shiva and the Quantum are tied with the greatest hard drive capacity: 40MB.
6. The Shiva has by far the best display quality, while the ICON is noticeably poor.
7. The ICON is by far the least expensive of the three machines at $700 advertised.
8. Although the Shiva is an impressive machine, its cost is out of our range. Its excellent screen quality, which probably accounts for its cost, is not currently a requirement for Software Designs.
9. Although poor in comparison to the other machines, the screen quality of the ICON is good enough for our employees' needs.
10. While the Quantum and the ICON are more less equal in terms of memory and battery life, the weight comparison is very favorable for the ICON.
11. Because it is the least expensive and the lightest of the machines, and because its other characteristics are acceptable, the ICON S/6000 is the best choice.

Recommendations

Based on these comparisons, the ICON S/6000 is the best laptop computer of the models we compared. It fulfills Software Design's basic requirements. However, we do urge further consideration of machines with high-quality graphic and color display capabilities. Machines like these would enable our products to make a much better impression on potential customers.

7

Figure 5-10 More key elements of a recommendation report: the conclusions section and the final recommendations.

Figure 5-10 briefly mentions which considerations were most influential in reaching the final recommendation.

Sometimes, recommendations cannot be so clear cut. Sometimes qualifications must be stated: for example, "if we want to control costs, then choose the ICON S/6000"; but "if we want high quality and function, then choose the Shiva HB60/S." In some situations, you may be forced to recommend none of the options. Imagine that you were asked to study the market for grammar- and style-checking software. You go out and investigate as many of the different packages as you can, reading the reviews on them, and getting as many product demos as you can. At the end, you throw up your hands, exclaiming that they are all worthless and that the company would be better off to hire a human being to do the copyediting.

GRAPHICS IN RECOMMENDATION REPORTS

As you plan a recommendation report, consider what sorts of illustrations, drawings, tables, or charts might be necessary. In this type of report, tables of comparative data are often necessary, as is shown in Figure 5-9. For more dramatic demonstrations of your points, you can use line graphs, pie charts, bar charts, and other such ways of depicting data.

EXERCISES

This chapter covers seven common types of engineering reports. Interview at least three professional engineers about the reports they write, and ask them questions like the following:

1. Which of these types of report do they most commonly write? Are there other types, not covered in this chapter, that they also write?
2. What are the chief purposes of the reports they write? Are the reports for internal or external consumption, for colleagues or clients?
3. How important are these reports to their business and professional careers? How much, for example, do they rely on proposals to get contracts? How often does the entire work product of their professional consultation take the form of a recommendation report?
4. Do they get editorial or production assistance in preparing these reports? Or do they handle the entire writing, editing, printing, and binding of their reports themselves?
5. Ask your engineering interviewees about the progress reports they write. What sorts of projects require progress reports? How often do they submit progress reports for a typical project?

BIBLIOGRAPHY

The following books and articles supply additional information on report-writing in general as well as on the types of reports covered in the chapter:

Cavin, Janis I. *Understanding the Federal Proposal Review Process.* Washington, DC: American Association of State Colleges and Universities, 1984.

Fischer, Martin A. *Engineering Specifications Writing Guide.* Englewood Cliffs: Prentice-Hall, 1983.

Hartley, J. *Designing Instructional Text.* London: Kogan Page, 1985.

Hill, James W., and Timothy Wahlen. *How to Create and Present Successful Government Proposals.* New York: IEEE Press, 1993.

Stewart, Rodney D., and Anne L. Stewart. *Proposal Preparation.* New York: Wiley, 1992.

6

WRITING AN ENGINEERING REPORT

As an engineer, you may often become involved in projects for which you must write reports. Engineering reports have *specifications* just as do the other sorts of projects you work on. Specifications for reports involve the layout of the report, the organization and content of the sections, the format of the headings and the lists, the design of the graphics, and so on. In fact, the American National Standards Institute (ANSI) has defined and published specifications for engineering reports entitled *Scientific and Technical Reports: Organization, Preparation, and Production*. The following discussion is based on those guidelines.

The advantage of having such a required structure and format is that when you or any other professional engineer picks up a report, it will be designed in a familiar way—you will know what to look for and where to look for it. Reports are usually read in a hurry—people are in a hurry to get to the information they need, the key facts, the conclusions, and other such essentials. That is another reason why there is a standard format for reports.

When you write your first engineering report, you may be struck by how repetitive some of the sections seem. The apparent duplication has to do with how professional engineers read reports. They don't read reports straight through: they may start in the middle, skip around, and probably not read every page. That means your challenge is to design reports so that your readers will encounter your key facts and conclusions, no matter how much of your report they read or in what order they read it.

The standard sections of the engineering report are as follows:

Transmittal letter
Covers and label
Title page
Table of contents
List of figures
Executive summary
Introduction
Body of the report
Appendixes (including references)

The following sections guide you through each of these standard sections, pointing out the key features. As you read and use these guidelines, remember that these are guidelines, not iron-clad laws. The standard for engineering reports is not intended as a straitjacket, but as a focal point to enable writers in the profession to maintain a familiar "look and feel" to their documents. You'll notice in your career that different companies, professions, and organizations have their own standards for reports—you'll need to adapt your practice to these as well.

LETTER OF TRANSMITTAL

The transmittal letter is a cover letter. An example is shown in Figure 6-1. Usually, it is attached to the outside of the report with a paper clip. It is a communication from you—the report writer—to the recipient, the person who requested the report and (who knows?) may be paying you for your expert consultation. Basically, it says "Okay, here's the report that we agreed I'd complete by such-and-such a date. Briefly, it contains this and that, but does not cover this or that. Let me know if it is acceptable."

The transmittal letter explains the context—the events that brought the report about. It contains information *about* the report that does not belong *in* the report.

In Figure 6-1, notice the standard business-letter format. If you write a report internally, you'll use the memorandum format instead; in either case, the contents and organization will be the same. Notice the contents of the letter:

- The first paragraph cites the name of the report, putting it in italics (if italics is available), underscores, or all caps. It also mentions the date of the agreement to write the report.
- The middle paragraph focuses on the purpose of the report and gives a brief overview of its contents.

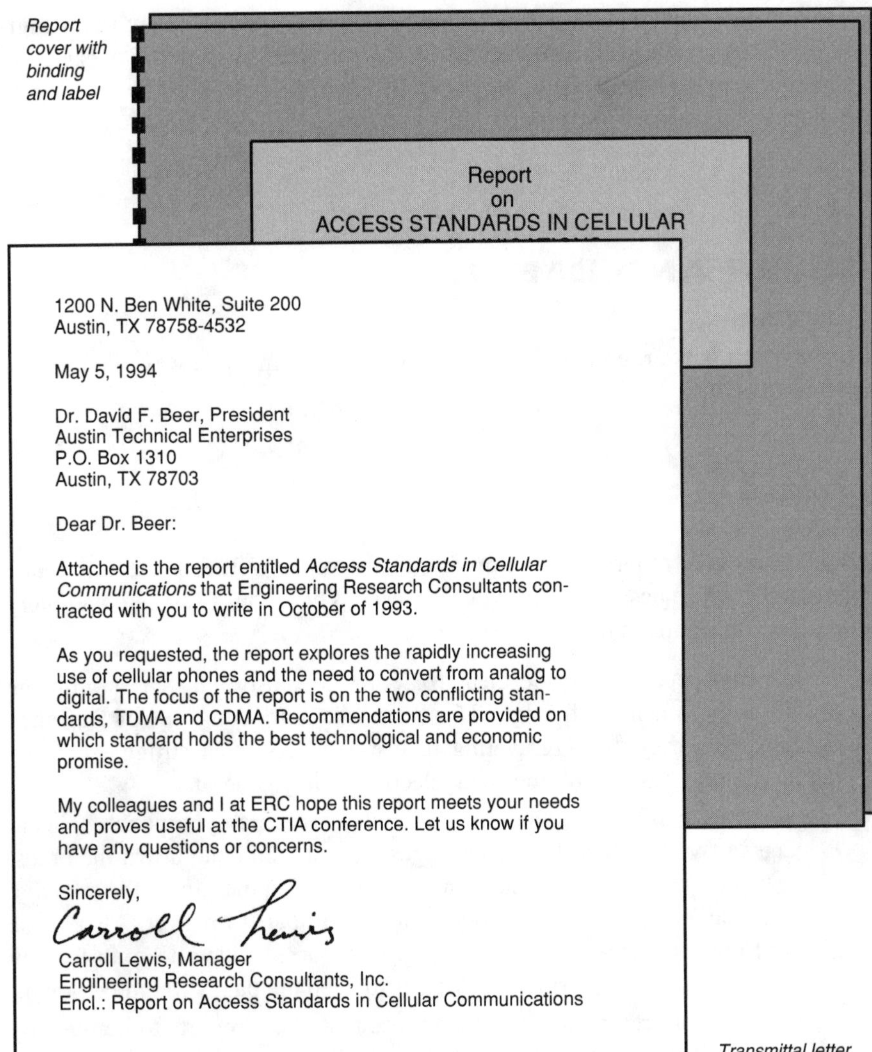

1200 N. Ben White, Suite 200
Austin, TX 78758-4532

May 5, 1994

Dr. David F. Beer, President
Austin Technical Enterprises
P.O. Box 1310
Austin, TX 78703

Dear Dr. Beer:

Attached is the report entitled *Access Standards in Cellular Communications* that Engineering Research Consultants contracted with you to write in October of 1993.

As you requested, the report explores the rapidly increasing use of cellular phones and the need to convert from analog to digital. The focus of the report is on the two conflicting standards, TDMA and CDMA. Recommendations are provided on which standard holds the best technological and economic promise.

My colleagues and I at ERC hope this report meets your needs and proves useful at the CTIA conference. Let us know if you have any questions or concerns.

Sincerely,

Carroll Lewis

Carroll Lewis, Manager
Engineering Research Consultants, Inc.
Encl.: Report on Access Standards in Cellular Communications

Figure 6-1 A cover letter and bound report with label on the front cover. Normally the transmittal letter is paperclipped to the front cover of the report.[8]

- The final paragraph encourages the reader to get in touch if there are questions, comments, or concerns. It closes with a gesture of good will, expressing hope that the reader finds the report satisfactory.

[8] Example material drawn from work done by Eric Manna and Jiandong Zhu, engineering students, University of Texas at Austin, 4 May, 1994.

Of course, the contents of this letter, as with any other element in an engineering report, may need to be modified for specific situations. For example, you might want to add another paragraph, listing questions you'd like readers to consider as they review the report.

COVER AND LABEL

If your report is over ten pages, you'll want to bind it in some way and create a label for the cover.

COVERS

Good covers give reports a solid, professional look as well as protecting them. There are many types of covers you can use. When you go to the stationery store, keep these tips in mind:

- Totally unacceptable are the clear (or colored) plastic slip cases with the plastic sleeve on the left edge. These are like something out of freshman English; plus they are aggravating to use—readers must struggle to keep them open and hassle with the static electricity they generate.
- Marginally acceptable are the kinds of covers for which you punch holes in the pages of your report, load your pages in, and then bend down the brads. These work, but remember to leave an extra half-inch margin on the left edge to keep readers from having to pry your report apart to read it. Also, these kinds of covers typically do not lie flat; they force readers to grab for any available object or use various body parts to keep the pages weighted down.
- By far the most preferable covers are those that allow reports to lie open by themselves. It's a great relief for a report to lie open in your lap or on your desk. Check with your local copy shop or stationery store; these sorts of bindings are inexpensive. Most of them use a plastic spiral for the binding and thick, card-stock paper that comes in a range of colors for the covers.
- Generally less preferable are loose-leaf notebooks, or ring binders. These are too bulky for short reports. Of course, the ring binder makes changing pages easy; if that's how your report will be used, then it's a good choice. Otherwise, it's a nuisance; it's bulky; and the page holes tend to tear.
- At the "high end" are those overly fancy covers with their leatherette look and gold-colored trim. Avoid them. Keep it plain, simple, and functional.

LABELS

Be sure to devise a label for the cover of your report. It's a step that some report writers forget. But without a label, a report will be ignored.

One common way to create labels is to type the report title and other essential information on an adhesive label and affix it to the cover. Often, these labels dry and fall off or don't adhere well to glossy surfaces.

A much cleaner way to create covers is to design one on a typewriter or in a computer file, print it out, then go to a copy shop and have it photocopied directly onto the report cover.

Not much can go on the label: the report title, your name, your organization's name, a report tracking number, and a date. There are no standard requirements for the label, although there might be in your specific company or organization. (An example of a report label is shown in Figure 6-1.)

TITLE PAGE

The title page should contain the following elements:

- The full title of your report
- The name, title, and organization of the individual you are submitting the report to
- Your name, title, and organization
- The date the report is submitted
- A descriptive abstract (included on the title page of some reports).

The example title page in Figure 6-2 illustrates most of these elements and their arrangement. Naturally, other elements may be required on the title page of certain reports. If it's part of a contract, the contract number may be required. If the report has a tracking number, that number may be needed both on the title page as well as on the cover label.

As with so many other aspects of professional life, when you work for a specific organization, you must be aware of how things are done in that organization—things that may not be written down anywhere but that are expected nonetheless. Take a look at a few of the reports where you work, and make a mental "spec sheet" on how they are designed.

Notice the layout of the elements on a title page—whichever ones are used. Most are centered and are carefully spaced vertically on the page so that a nice, roomy design is achieved. You don't want to cram all of the elements at the top, bottom, or middle of the title page; spread them out over the whole vertical

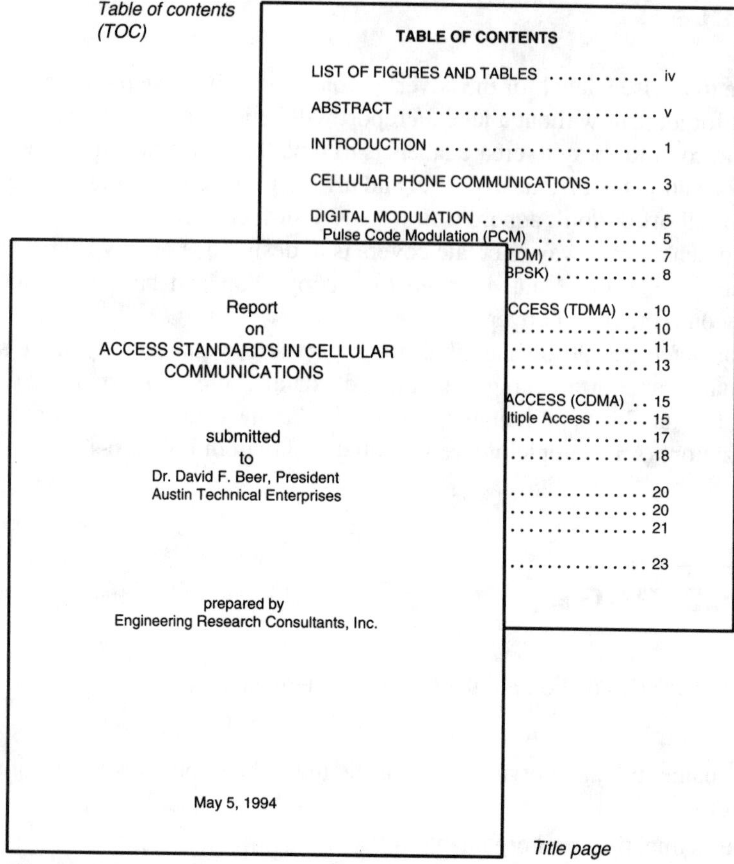

TABLE OF CONTENTS

LIST OF FIGURES AND TABLES iv
ABSTRACT v
INTRODUCTION 1
CELLULAR PHONE COMMUNICATIONS 3
DIGITAL MODULATION 5
Pulse Code Modulation (PCM) 5
TDM) 7
BPSK) 8
CCESS (TDMA) ... 10
.... 10
.... 11
.... 13
ACCESS (CDMA) .. 15
ltiple Access 15
.... 17
.... 18
.... 20
.... 20
.... 21
.... 23

Report
on
ACCESS STANDARDS IN CELLULAR COMMUNICATIONS

submitted
to
Dr. David F. Beer, President
Austin Technical Enterprises

prepared by
Engineering Research Consultants, Inc.

May 5, 1994

Figure 6-2 A title page and table of contents from an engineering report. (Some reports include a descriptive abstract at the bottom of the title page.)

length of the page.

ABSTRACT AND EXECUTIVE SUMMARY

Most engineering reports contain at least one abstract—sometimes two—in which case the abstracts play slightly different roles. Abstracts summarize the contents of a report, but the different types do so in different ways.

One common type of abstract is the *descriptive* abstract. It provides an overview of the purpose and contents of the report. Another common type is the *executive summary*, which summarizes the key facts and conclusions contained in the report. The title page for some report designs contains a descriptive abstract; Figure 6-3 illustrates a typical executive summary.

The executive summary summarizes the key contents of the report. It's as if you used a yellow highlighter to mark the key sentences in the report and then copied them all onto a separate page and edited them for readability. Typically, executive summaries are one-tenth to one-twentieth the length of reports ten to fifty pages long. For longer reports, ones over fifty pages, the executive summary should not go over three doublespaced typewritten pages. The point of the executive summary is to provide a summary of the report—something that can

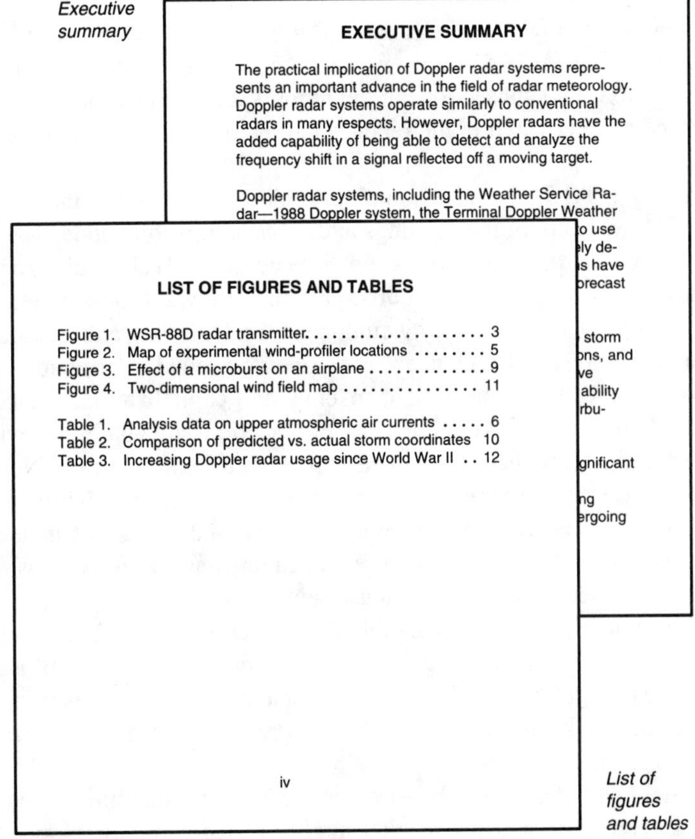

EXECUTIVE SUMMARY

The practical implication of Doppler radar systems represents an important advance in the field of radar meteorology. Doppler radar systems operate similarly to conventional radars in many respects. However, Doppler radars have the added capability of being able to detect and analyze the frequency shift in a signal reflected off a moving target.

Doppler radar systems, including the Weather Service Radar—1988 Doppler system, the Terminal Doppler Weather

LIST OF FIGURES AND TABLES

iv

Figure 6-3 A list of figures and the executive summary in an engineering report.

be read rather quickly.

If the executive summary, the introduction, and the transmittal letter strike you as repetitive, remember that readers don't necessarily start at the beginning of a report and read page by page to the end. They skip around: They may scan the table of contents to get a sense of the contents; they usually skim the executive summary for key facts and conclusions. They may read carefully only a section or two from the body of the report, and then skip the rest. For these reasons, reports are designed with some apparent duplication so that readers will be sure to see the important information no matter where they dip into the report.

TABLE OF CONTENTS

You're familiar with tables of contents (TOC) but may never have stopped to look at their design. A TOC shows readers the page number on which each of the major sections and subsections in the report starts. A TOC also shows readers what topics are covered in the report and how those topics are discussed (in other words, the subtopics).

In creating a TOC, you have a number of design decisions. One of the most important is how many of the headings and subheadings to include. The TOC is a collection point for the section titles and the headings and subheadings occurring within those sections. In longer reports, you may not want to include all of the lower-level headings because they would make the TOC long and unwieldy. The TOC should provide an at-a-glance way of finding needed information quickly.

Critical to a TOC is indentation, spacing, and capitalization. Notice in the example TOC in Figure 6-2 that the first-level sections are all aligned with each other; the second-level sections aligned with each other; and so on. Notice that page numbers are right-aligned with each other so that the last digit in a number is always in the same column. Notice also how capitalization is handled: Main chapters or sections are all caps; first-level headings are headline caps. If there were lower-level sections, they would use sentence-style caps.

Vertical spacing in a TOC is another design variable. Your goal is to spread the TOC nicely out on the page and to avoid just two or three lines of it spilling over to the next page. You can play with the spacing between lines to make this come out right, but keep it consistent. Make sure the spacing between headings of the same type is consistent.

One final note: Make sure the words in the TOC are the same as they are in the text. As you write and revise, you might change some of the headings—don't forget to go back and change the TOC accordingly.

LIST OF FIGURES AND TABLES

The list of figures has many of the same design considerations that the table of contents does. With the list of figures, the idea is to enable readers to find the illustrations, diagrams, tables, and charts in your report. The title shown in the list of figures is often shorter than it is in the actual text where the figure occurs. In the figure list, it's a good practice to shorten long figure titles to something complete and meaningful that readers can scan quickly.

Some complications arise when you have both tables and figures. Strictly speaking, *figures* are any illustration, drawing, photograph, graph, or chart. *Tables* are rows and columns of words and numbers and are not normally considered figures.

For longer reports that contain half a dozen or more of both figures *and* tables, you can create separate lists of figures and tables. Put them together on the same page if they fit, as shown in Figure 6-3. You can combine the two lists under the heading, "List of Figures and Tables," as in the figure.

INTRODUCTION

An essential element of any report is its introduction—make sure you are clear on its real purpose and contents. In an engineering report, the introduction prepares the reader to read the main body of the report. It does not dive into the technical subject, although it may provide a bit of theoretical or historical background. Instead, introductions indicate or discuss the following (but not necessarily in this order):

- Specific topic of the report (indicated somewhere in the first paragraph)
- Intended audience of the report; the knowledge or background that readers need to understand the report
- Situation that brought about the need for the report
- Purpose of the report—what it is intended to accomplish (as well as what it specifically does not intend to accomplish)
- Contents of the report—usually a numbered list of the key topics covered
- Scope of the report—what the report does not cover
- Background (such as concepts, definitions, history, statistics)—just enough to get readers interested, just enough to enable them to understand the context.

A common problem in writing introductions occurs when the discussion of background gets out of hand and runs on for several pages. For a typical twenty-page report, for example, the introduction shouldn't be too long—no more than two pages. You may view introductions as the place for discussing background. Ordinarily, that's not the case—the introduction prepares readers to read the report; it "introduces" them to the report. If there is just too much background to cover, move it to a section of its own, either just after the introduction or into an appendix.

Take a moment to look at the introduction in Figure 6-4. Notice how it handles the items in the preceding list. For some elements, the wording is straight-

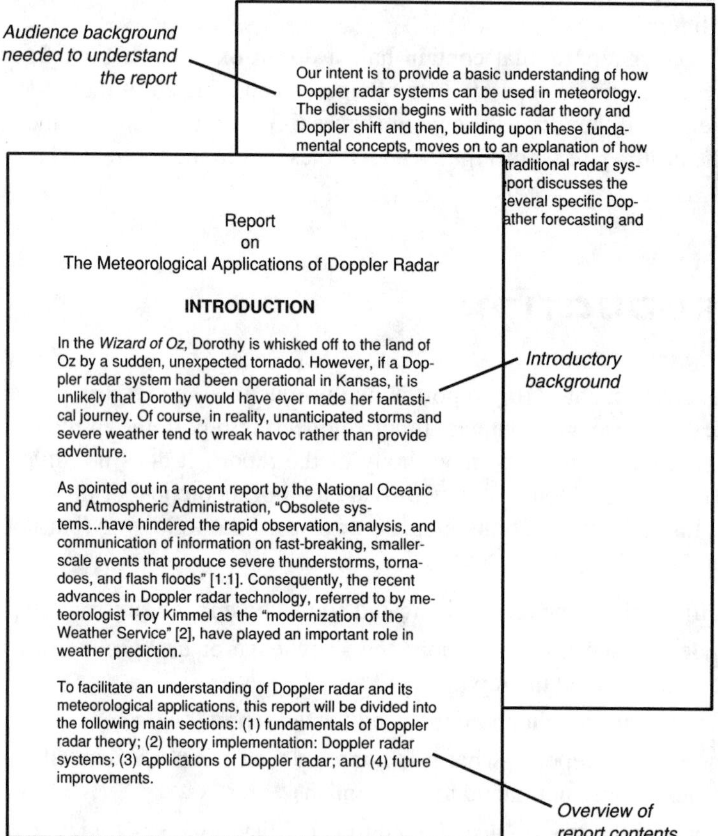

Figure 6-4 Introduction to an engineering report.[9]

[9] Example material drawn from work done by David Barron and Stephanie Braun, engineering students, University of Texas at Austin, 4 May, 1994.

forward ("This report will be divided into..."). Other elements are indicated subtly, as is the case with the background the audience needs to read the report: "a basic understanding" suggests that no specialized knowledge is needed to understand the report.

THE BODY OF THE REPORT

The body of the report is of course the main text of the report, the sections between the introduction and conclusion. Figure 6-5 shows a couple of sample pages.

HEADINGS

In all but the shortest reports (two pages or less), use headings to mark off the different topics and subtopics covered. This will enable readers to skim your report and dip down at those points where you present information that they want.

LISTS

In the body of your report, you'll also want to use the various kinds of lists where appropriate. Lists help readers by emphasizing key points, by making information easier to follow, and by breaking up solid walls of text. For example, if you have three key points that readers must not overlook, use a bulleted list. If you have a sequence of steps readers must perform, use a numbered list. If you have some key terms and definitions that need to stand out, use a two-column list.

SYMBOLS, NUMBERS, AND ABBREVIATIONS

Technical-report discussions often contain lots of symbols, numbers, and abbreviations. Remember that the rules for using numerals as opposed to words are different in the technical world. The old rule about writing out all numbers below 10 does not always apply in engineering reports.

SOURCES OF BORROWED INFORMATION

To write your report, you may have to borrow facts and ideas from other engineers as well as from people in other professions. When you do, you must

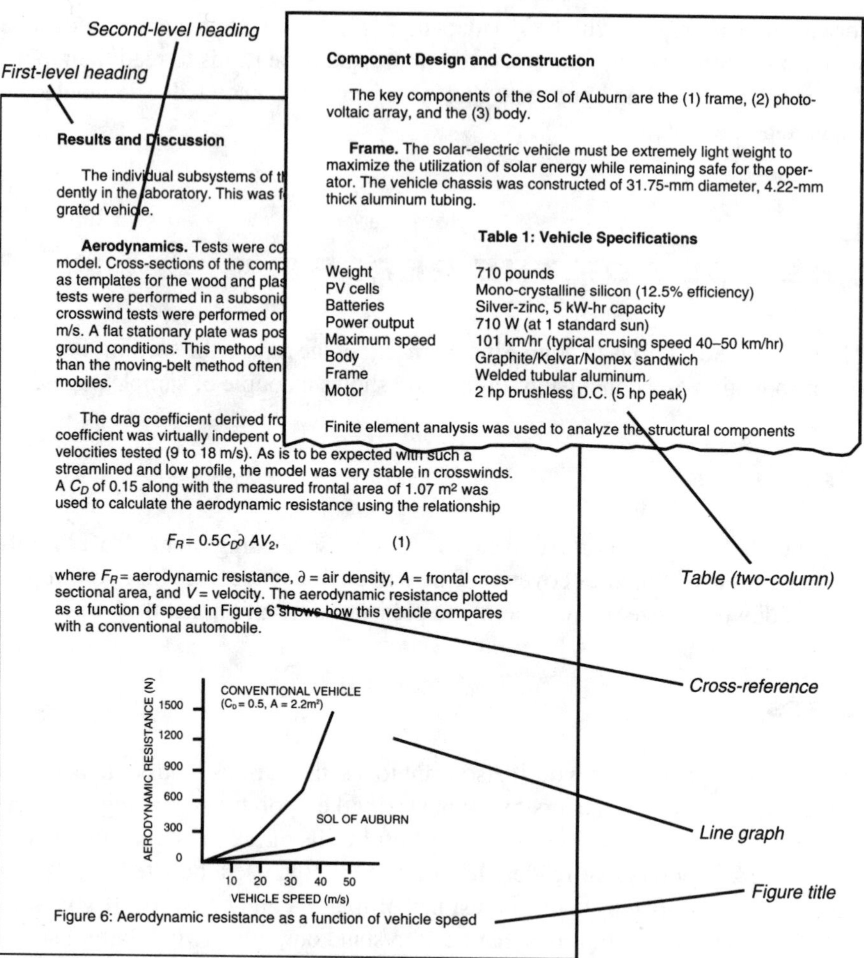

Component Design and Construction

The key components of the Sol of Auburn are the (1) frame, (2) photovoltaic array, and the (3) body.

Frame. The solar-electric vehicle must be extremely light weight to maximize the utilization of solar energy while remaining safe for the operator. The vehicle chassis was constructed of 31.75-mm diameter, 4.22-mm thick aluminum tubing.

Table 1: Vehicle Specifications

Weight	710 pounds
PV cells	Mono-crystalline silicon (12.5% efficiency)
Batteries	Silver-zinc, 5 kW-hr capacity
Power output	710 W (at 1 standard sun)
Maximum speed	101 km/hr (typical crusing speed 40–50 km/hr)
Body	Graphite/Kelvar/Nomex sandwich
Frame	Welded tubular aluminum.
Motor	2 hp brushless D.C. (5 hp peak)

Finite element analysis was used to analyze the structural components

Results and Discussion

The individual subsystems of t
dently in the laboratory. This was f
grated vehicle.

Aerodynamics. Tests were co
model. Cross-sections of the comp
as templates for the wood and plas
tests were performed in a subsonic
crosswind tests were performed or
m/s. A flat stationary plate was pos
ground conditions. This method us
than the moving-belt method often
mobiles.

The drag coefficient derived fro
coefficient was virtually independent o
velocities tested (9 to 18 m/s). As is to be expected with such a streamlined and low profile, the model was very stable in crosswinds. A C_D of 0.15 along with the measured frontal area of 1.07 m² was used to calculate the aerodynamic resistance using the relationship

$$F_R = 0.5 C_D \partial A V_2, \qquad (1)$$

where F_R = aerodynamic resistance, ∂ = air density, A = frontal cross-sectional area, and V = velocity. The aerodynamic resistance plotted as a function of speed in Figure 6 shows how this vehicle compares with a conventional automobile.

Figure 6: Aerodynamic resistance as a function of vehicle speed

Figure 6-5 Pages from the body of an engineering report. Note the use of headings, tables, cross-references, and graphics.

indicate the sources of your borrowed information, which is known as *documenting* your sources.

GRAPHICS AND FIGURE TITLES

In your engineering report, you're likely to need drawings, diagrams, tables, and charts. These not only convey certain kinds of information more effectively but also give your report an added look of professionalism and authority. If you've never put these kinds of graphics into a report, there are some relatively easy ways to do so—you don't need to be a professional graphic artist!

CROSS-REFERENCES

In your reports, you may need to point readers to other places in the report where closely related information is discussed, or to other books and reports that have useful information. These are called *cross-references*. For example, they can point readers from the discussion of a mechanism to an illustration of it. They can point readers to an appendix where background on a topic is given (background that just does not fit in the text). And they can point outside your report to other information—to articles, reports, and books that contain information related to yours. When you create cross-references, follow these guidelines:

- If you refer readers to another section of your report, put the heading or section title in quotation marks.
- If you refer to the title of a journal, book, or report, underline or italicize that title.
- If you refer to an article in a journal or encyclopedia, put quotation marks around the article title.

When you create cross-references, give readers some clue as to why they should see that information. Otherwise, they are likely to wonder. Indicate the topic of the cross-referenced information (don't assume the title indicates it fully) and suggest why readers might want to follow the cross-reference.

There are no rules as such on when to cite exact titles or when to supply page numbers in cross-references. The guiding principle is to make it easy on the reader. In a short report, say, one under twenty pages, citing page numbers may not be necessary (although word-processing software makes it easy to automate these details). If you supply the page number, then you can cite the subject matter of the section, not the exact title. It's common to shift text around, cut sections, add sections, change the wording of headings—all of which plays havoc with your cross-references.

CLARITY OF WRITING STYLE

As you rough-draft your report, don't get stymied over getting the words exactly right or avoiding grammar mistakes. In the rough-drafting stage, focus on the technical subject matter and don't get hung up on picky details that just slow you down.

However, once you've got a rough draft on paper or (more likely) in a computer file, reread it looking specifically for the common "writing style" problems that make engineering writing, or any writing, hard to read:

- *Unnecessary passive voice:* In the technical world, you may have to use the passive voice, but when it is misused, it leads to unclear, wordy writing.
- *Over-reliance on the* be *verb:* Heavy use of the *be* verb can make writing unclear and wordy as well.
- *Unnecessary expletives:* The expletives we mean here are the ones that use some form of "it is" or "there is." They too can inflate writing, making it less direct and understandable.
- *Redundant phrasing:* For examples of wordy phrases and their concise counterparts, see the section on redundancy in Chapter 3.
- *Noun stacks:* Another problem, particularly in the technical world, involves jamming three or more nouns together into a phrase, called a noun stack.
- *Weird combinations of subjects and verbs:* When you are struggling to express complex technical ideas, it's easy to combine subjects and verbs in strange ways, especially when lots of words come between them in the sentence. For example:

> The *causes* of the disappearance of the early electric automobiles *were devastating* to the future of energy conservation in the United States.

  In this example, it should be the "disappearance" that was "devastating," not the "causes."

- *Sentence length:* In the discussion of complex technical matters, longer sentences may be necessary; but review them to see whether splitting them would make for easier comprehension.

All of these strategies for clearer, more economical writing are discussed in detail in Chapter 3, "Eliminating Sporadic Noise in Writing."

Paragraph Structure

When you review your rough draft, look for ways to strengthen the organization and flow of your ideas. Do this kind of review at the level of whole paragraphs and whole groups of paragraphs:

- Strengthen transitions between major blocks of thought, such as between paragraphs or groups of paragraphs.

- Add topic sentences (particularly the overview kind) to paragraphs where appropriate.
- Check the logic and sequence of paragraphs or groups of paragraphs. To do so, label each paragraph or paragraph group with one or two identifying words—this way you can get the "global picture" more easily.
- Break paragraphs that go on too long, challenging the reader's attention span.
- Consolidate clusters of short paragraphs that focus on essentially the same topic. Too many paragraph breaks can have a fragmented and distracting effect.
- Interject short overview paragraphs at the beginning of sections and subsections.

Using these strategies guides readers through your report, showing them what lies ahead, where they have come from in the previous pages of the report, and how everything fits together.

GRAMMAR, USAGE, AND PUNCTUATION

As with writing style, you don't want to slow yourself down worrying about subjects and verbs, commas, apostrophes, and the like. Worry about these details later. However, once you've got a rough draft on paper or on disk, check for the various common mistakes such as those involving commas, apostrophes, spelling (particularly spelling of similar-sounding words), parallelism, agreement, and so on.

See Chapter 3, "Eliminating Written Noise," for details on grammar, punctuation, and usage rules.

PAGE NUMBERING

At first glance, the style for numbering pages in a report may seem arcane. But it is based on traditional publishing practice and can be reduced to a few simple rules:

- All pages in the report are numbered; but on some pages, the numbers are not displayed.
- All pages before the first page of the introduction use lowercase Roman numerals; all pages beginning with and following the introduction use Arabic numerals.

- Longer reports often use the page-numbering style known as folio-by-chapter or double-enumeration (for example, pages in Chapter 2 would be numbered 2-1, 2-2, 2-3, and so on). This style eases the process of adding and deleting pages and enables readers to know where they are in the report.
- On special pages, such as the title page and page one of the introduction, the page number is *not* displayed. (Imagine how a lowercase roman numeral "i" at the bottom of a title page would look.)
- Page numbers can be placed in one of several areas on the page, but wherever you place them, do so consistently. If you are printing or typing your report single sided, the best and easiest choice is to place page numbers at the bottom center of the page (and to hide them on special pages).
- If you place page numbers at the top of the page, you must hide them on chapter or section openers where a heading or title is at the top of the page. (Again, imagine how this would look.)

GRAPHICS

When you write your report, you're likely to need illustrations, diagrams, charts, graphs, drawings, schematics, and tables. Graphics like these help present your information more effectively and give a polished, professional look to your report. You don't need to be a graphics professional to bring good graphics into your engineering reports. At the very minimum, all you need is some scissors or an X-acto knife, some tape or glue, and access to a good photocopying machine.

However, there are many advantages to using the computer-based graphics tools that enable you to create or scan graphics. And of course with these tools, you need word-processing software that enables you to embed those graphics into your text files. These software approaches for incorporating graphics save you the trouble of physically cutting and pasting graphics into the pages of your report. These tools also enable you to send text and graphics files electronically to professional associates and to publishers. Thus, it's worth spending some time getting comfortable with handling graphics electronically. For an overview of features to look for in computer-based graphics tools, see Chapter 10.

AN OVERVIEW OF GRAPHICS

If you're new to using graphics in reports, consider the sorts of graphics you can use:

- *Drawings:* Drawings are simplified illustrations of objects, people, and places. You often see drawings used in instructions. They strip away extraneous detail and focus on the key subject matter.
- *Photographs:* Photographs, on the other hand, supply lots of detail—in some cases, too much. They are useful, for example, when you want to show a model of a new product.
- *Diagrams and schematics:* Diagrams are highly abstract illustrations of objects. They often focus on infrastructural matters such as circuitry and other detail. They are often accompanied by measurements and symbols.
- *Conceptual diagrams:* Graphics are also used to illustrate nonphysical things such as concepts. An organizational chart of a company is a typical example. A flowchart of a production process is another.

DESIGN AND FORMAT OF GRAPHICS

When you incorporate graphics into a report, pay attention to their standard components, their placement, and cross-references to them.

Components of Graphics. When you use graphics, keep these design considerations in mind:

- Add *labels*—Words that identify the parts of the thing being illustrated, and a pointer from each label to the part being illustrated.
- Add *figure titles*—Those identifying titles at the bottom of figures that indicate the subject matter of each figure and its sequence number in your report.
- Place graphics at the *point of first reference*—Position graphics just after the first point in your text where they are referenced, if not on the same page, then at the top of the next.
- *Intersperse* graphics with text—Insert graphics into pages with text rather than appending them at the end of the report. For readers, it's pleasing to have text broken up with graphics.
- Provide *cross-references* to your graphics—Don't just pitch graphics into your report without referring to them and explaining key points about them.

These components enable a graphic to stand by itself and make sense. The figure title should provide only a brief statement of the graphic's significance;

the textual discussion of the graphic can expand upon this more fully. Remember, the purpose of a graphic is to help your reader understand the information presented in your text, but the text must do the same for the graphic. That is, your reader needs to know, from your text, what to look for in the graphic.

Placement of Graphics. Each graphic should appear as soon as possible after you first mention it and as close as possible to your discussion of it. If there is no room between first mentioning it and the end of the page, put it on the next page but tell readers where it is with wording such as "As shown in Figure 6 on page...." Make sure there is adequate spacing between the graphic and the text and that the page is visually attractive and balanced.

The way to merge a graphic with your text depends on its shape and size. Graphics are usually full-page, half-page, or smaller than half-page. In each case, there are some general guidelines to follow:

- If you have a full-page graphic and if your document has left and right pages, place the graphic on the right-hand page if at all possible. Your discussion of it can then appear first on the left-hand page; both the graphic and your discussion of it will be visible.
- If you write a report with only one full-page graphic in it, follow the previous guidelines unless you refer to the graphic throughout the report. Then it's better to place it in an appendix at the end of the report. Be sure to cite its location each time you mention it—people forget.
- If your graphic is horizontal rather than vertical, make sure its top is placed along the inside of the page. For a bound report or book, this means placing the top side against the spine.
- If you have a half-page graphic that won't fit on the page where you begin discussing it, put it at the top of the following page.
- Many smaller graphics really should be a half-page in size. Don't be tempted to economize on space at the cost of readability. If you include wording within a smaller graphic, the print needs to be large enough to read. Make sure you don't challenge the eyesight of your readers—and thus create noise—when you provide them with a small graphic.
- If you create a graphic less than a half-page in size, you can have your text flow around it (Figure 6-6*b*). This is easy enough to do with modern word processing programs and gives your page a unique and professional appearance. Don't cramp things, however. Make sure you leave plenty of white space between your text and graphic.

Cross-References to Graphics. As indicated above, if you don't refer to the graphic, your reader may be left with a nice picture but no sense of its purpose

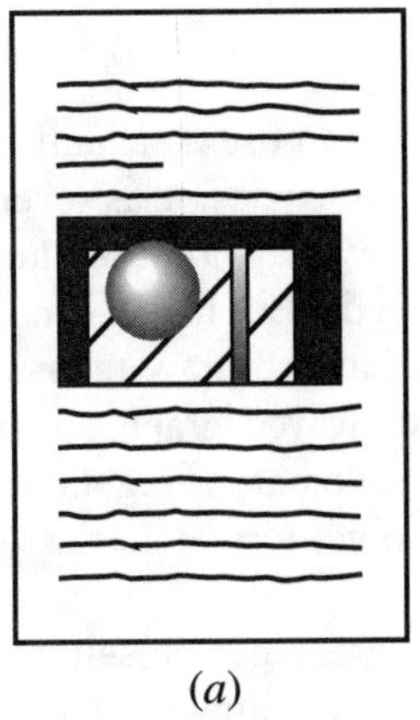

(*a*)

(*b*)

Figure 6-6 (*a*) Example of effective centering of a graphic on the page. (*b*) Letting text flow around a graphic can give your page a professional look.

or meaning. Moreover, if your first reference to and discussion of a graphic come after it has already appeared, your reader may be reluctant to turn back to it. Always refer to your graphic and begin your discussion of it before it appears in your text. You can do this with phrasing like the following:

> As can be seen in Figure 5, the thermophysical properties...
>
> The arrangement of the MOF network (Fig. 8.2) is structured so as to...
>
> The averages for the fabric cutting speeds are shown in Table 4 on the next page.

The word Figure is usually capitalized, but it can be abbreviated as Fig. There is no abbreviation for Table.

SOFTWARE APPROACHES TO GRAPHICS

As mentioned previously, there are advantages to using graphics and word-processing software that enable you to integrate graphics and text. You don't have to tape graphics in, and you can send your reports to colleagues and publishers electronically.

By now, most of the leading commercial providers of word-processing software enable you to "embed" graphics into text files. WordPerfect, Microsoft

Word, AmiPro, and others provide this function, but of course the techniques are dramatically different with each software package.

One approach to getting graphics into reports is to use a scanning device. Scanning equipment can be as cheap as $50.00 and as expensive as several thousand dollars. Watch out for the low-end scanners; they may produce blurry, low-quality images and may not be suitable for your needs. But if you can get good scanned images, it's a big advantage—you don't have to draw them yourself. Remember, however, you still must document borrowed graphic images the same as you must document borrowed text. And, if you are using the borrowed graphic in a commercial publication, you must also get formal permission from the originator of the graphic.

The other alternative is electronically drawing graphics yourself using software tools such as CorelDraw, AutoCAD, or any of the drawing tools that are available with most desktop-publishing systems.

"Low-Tech" Graphics Production

If you don't have a scanner, have no access to or skill with software graphics tools, and consider yourself a horrible artist, there's still hope. Try the following technique:

1. Find the graphic that you want in a book, report, or journal, for example. Avoid graphics in low-quality print media such as newspapers; they won't photocopy well.
2. Photocopy it on a good-quality copier. Enlarge or reduce the image as needed.
3. Trim the copy, cutting out the figure title, wordy legends, but not necessarily the labels (usually they will work just fine as your own labels).
4. Add labels or other devices as necessary to make the graphics work in your report.
5. In the text of your report, plan where you'll place the graphic: try for the point at which it is first referred to or some point just after (to make the pagination work). Leave enough space above and below the graphic so that it won't appear "squeezed" in. Make sure it doesn't spill outside your regular right and left margins.
6. In your file or on your typed page, type the figure title. Because you've photocopied the graphic, you must cite the source, just as with any information you borrow.
7. When you've printed out or typed your report, carefully place your graphic in the space you've left open for it, and then tape or glue it in in such a way that the seams and the tape won't show in the photocopy.

8. Now, get a good-quality photocopy not only of this page but of *all* the pages in your report. *Never* submit a report with things taped or stapled or clipped in the pages. (And by the way, don't draw or color on your graphics; if that's what you want, use a color photocopier.)

It may not be so out of the question for you to create some of your graphics yourself. Consider tracing the images you want. If you draw freehand, use a soft pencil and light marking to get the drawing just right, then ink it in with a black marker. Erase your pencillings, then treat your drawing just like the photocopied graphics discussed above.

CHARTS

The term *charts* encompasses all those ingenious ways of showing relationships between data—for example, line graphs, bar charts, pie charts, and three-dimensional variations of these such as pictographs. All of these types of charts are visual representations of tables. They express a fundamental frustration with the dull old table—row upon row and column upon column of numbers and words.

In tables, it is normally difficult to perceive what is significant about the data without studying it. Charts and graphs make the significant stand out. For example, if your department has reduced defects in the manufacturing process each year over the past five years, a line graph shows this more vividly than a table. If those defects are primarily the result of faulty raw materials, then a pie chart might make this point much more vividly than a table.

Obviously, charts and graphs are great ways to dramatize key statistical points, or "trends" in the data. But how do you construct them and then incorporate them into your reports? Most of the major word-processing and desktop-publishing software packages now have chart-making features. You feed in the data and define how the chart should look, and the program constructs it and embeds it where you want it in your report. Spreadsheet and database programs also can produce charts and graphs, which some word-processing and desktop-publishing software packages can import. And of course there are the well-known chart-making programs such as Harvard Graphics designed specifically for this function. Again, many word-processing and desktop-publishing software packages can import these charts.

And finally there are the manual approaches: You can use simple graphics capabilities in your word-processing software to draw your chart or graph. One other way is to find the chart or graph you want in some other published source, copy or scan it, and bring it in manually or electronically into your text. Yes, this is legal, as long as you document it.

Whichever means you use to create charts and then incorporate them into your report, observe a few fundamental guidelines:

- Include a figure title just below the chart to identify its content and, if necessary, to identify its source (if you borrowed the data).
- Add labels on the axes to identify the units of measurement.
- Include a "legend" if you use different symbols, colors, shadings, or patterns to indicate different elements.
- Make sure your charts fit within your regular margins.

TABLES

Despite what we've just said about tables compared to charts, tables are useful and necessary elements. Tables present data efficiently; an added benefit is that tables, like charts, lists, graphics, and headings, break up big walls of text. Report writers often pass up good opportunities to present data in tabular form and instead leave it in regular paragraphs. Any time you see groups of numbers in your text, take a second look at them to see if they can be presented, or re-presented in the form of tables. Figure 6-7 illustrates how you can transform text into a table.

The sources of your data for tables can be varied. It's perfectly legal to copy a table from another source into your own report, as long as you document its origins. How you construct the table though is another matter. As with charts and graphs, there are numerous software tools such as Harvard Graphics that you can use to define the table you want, feed in your data, and then embed the results into your text. There is the "low-tech" method of photocopying tables from other sources and taping them in your text. And still another low-tech method involves hand-typing tables yourself.

Whichever technique you use to create tables, keep these design considerations in mind:

- Include a heading at the top of each column to identify the contents of the column.
- Include a row heading in the farthest left column to identify the contents of the row.
- Center columns of numbers under column headings; left-justify columns of text.
- Right-align or decimal-align columns of numbers.
- Don't forget to indicate units of measurement; and put these units in the headings rather than by each item in the columns.

For many customers, the size, weight, and other physical aspects of a printer are important in their purchasing decision. The physical dimensions of the three printers are as follows. The Morton is 8.3 inches in height, 16.1 inches in width, and 15.4 inches in depth, and weighs 36 pounds. The IQ is 10.2 inches in height, 24.9 inches in width, and 16.0 inches in depth, and weighs 25 pounds. The Overture is 10.2 inches in height, 24.9 inches in width, and 16.0 inches in depth, and weighs 23 pounds.

Original version: data presented in regular paragraph form.

For many customers, the size, weight, and other physical aspects of a printer are important in their purchasing decision. The physical dimensions of the three printers are as follows:

Printer	Height (inches)	Width (inches)	Depth (inches)	Weight (pounds)
Morton	8.3	16.1	15.4	36
IQ	10.2	24.9	16.0	25
Overture	10.2	24.9	16.0	23

Revision: data now presented in a table.

Figure 6-7 Transforming text into table. In the original example, the data clogs up the textual discussion; in the revision, it is taken out of paragraph form and put into a table, making it more readily scannable and breaking up the text.

CONCLUSIONS

For most reports, you'll need to include a final section, usually called a "conclusion." When you plan and write final sections of engineering reports, think about the functions they can perform in relation to the rest of the report:

- They *conclude*—that is, they draw logical conclusions from the discussion that has preceded; they make inferences upon what has preceded.
- They *summarize*—that is, they review the key points, key facts, and so on from what has been discussed. Summaries present nothing new—they leave the reader with a perspective on what has been discussed, the perspective that the writer wants them to have.
- And finally they *generalize* by moving away from the specific topic to a discussion of such things as implications, applications, and future developments—but only in general terms.

Your final section can do any combination of these, depending on your sense of what your report needs. See the example conclusion in Figure 6-8; it

An appendix to the report

APPENDIX A: FUTURE DEVELOPMENTS

The intent of this report is to review current implementations of Doppler radar. However, important activities are going on currently to improve this technology.

Although Doppler radar has proven to be an important step in the future of radar meteorology, current Doppler radar systems are not perfect. The main disadvantage of a single Doppler radar system is that only the *radial* component of wind velocities can be detected. This means that only the vector component of the wind that is blowing directly toward or *away* from the ... sured. The obvious problem ... usly strong winds will not ... ficant portion of their velocity ... the system antenna [2].

... n the ... ation. ... e ... ar ... to ... sional ... Dual- ... that ... parate- ... e data ... specific ... ng

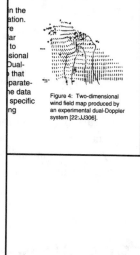

Figure 4: Two-dimensional wind field map produced by an experimental dual-Doppler system [22:JJ306].

CONCLUSION

Although basic Doppler radar systems have been in use since World War II, only recent developments in computer technology have enabled the large quantity of data collected by operational systems to be processed using state-of-the-art computer algorithms. The versatility of both hardware and software components of system computers make Doppler radar systems feasible for practical use. Each of the Doppler radar systems discussed in this report is still in the developmental stages; however, each has shown marked advantages over conventional radar systems when applied to weather forecasting and aviation meteorology. In addition, new dual-Doppler systems are currently being developed to increase weather forecasting accuracy by detecting otherwise hidden atmospheric conditions.

The implementation of the Weather Service Radar, Doppler, Terminal Doppler Weather Radar, and wind-profiling systems were important steps in the modernization of the National Weather Service. Each of these systems provides meteorologists with accurate information on current weather conditions, allowing them to draw conclusions about long-term weather forecasts. More importantly, meteorologists are able to predict and detect dangerous situations, such as flooding, tornadoes, microbursts, and wind shear, which pose a threat to the general public, as well as to the aviation industry.

11

Conclusion to the report

Figure 6-8 Conclusion section and appendix of an engineering report. Note that each is a separate section and begins on a new page.

summarizes key points in the report and takes a brief look at the future of the technology discussed in the body of the report.

The length of the conclusion can be anything from a 100-word paragraph to a five- or six-page section. For the typical ten- to twenty-page double-spaced report, the final section would be one to two pages, but such ratios should never be applied without considering what's going on in the report. Watch out for conclusions that get out of hand and become too long. Readers expect a sense of closure, a feeling that the report is ending. When the final section becomes too long, consider doing one of the following: Move some of the discussion back into the body of the report; shorten and generalize the discussion and keep it in the conclusion; or find some other way to end the report.

APPENDIXES

Appendixes are those extra sections following the conclusion. What sorts of things do you put in appendixes?—anything that does not comfortably fit in the main part of the report but cannot be left out of the report altogether. The appendix is commonly used for large tables of data, big chunks of sample code, foldout maps, or large illustrations that just do not fit in the body of the report. Anything that you feel is too large for the main part of the report or that you think would be distracting and interrupt the flow of the report is a good candidate for an appendix. Figure 6-8 shows an example of an appendix.

DOCUMENTATION

Documentation is the system by which you indicate the sources of the information you borrow in order to write an engineering report. Back in the "old days," they called it "footnoting"—thank goodness there are easier documentation systems to use (and word-processing software, which also makes footnoting much easier). As you probably know, writers document their information borrowings in order to

- enable readers to track down the information so that they can read it for themselves.
- protect the originator, the author of the information, so that she or he will get the credit and acknowledgment for having developed it.
- protect you from accusations of plagiarism—of stealing other people's hard-fought information discoveries.
- demonstrate to readers that you have done your homework, that you are "up" on the latest developments in this particular field.

Documenting your information sources has a lot to do with establishing, maintaining, and protecting your credibility in the profession. Borrowed information must be documented regardless of the shape or form in which it is presented; whether you directly quote it, paraphrase it, or summarize it, it's still *borrowed* information.

Nearly every field and profession has its own documentation system. A *documentation system* is the style and format that a particular field or profession uses. Even though systems for documenting sources of borrowed information vary, they can be reduced to a few basic categories. The system used by professional engineers in their publications and recommended by the IEEE fits into the category of *number systems*. As Figure 6-9 illustrates, at the end of the report there is the *references page*, a numbered list of sources; in the body of the report there are *textual references*, codes that work with the references page to indicate the source of the borrowed information. Readers understand that if they want to check out your sources on a topic, they can look up the code number in the back of the document to see where the information came from.

REFERENCES PAGE

At the back of the report is the list of information sources, arranged and numbered according to their occurrence in the report. For example, if the first borrowed information that occurs in your report comes from a book by Robertson, Robertson would be [1]. If the next borrowed information came from an article by Adams, Adams would be [2]. But once you've established an information source in the references page, no further entries for it are needed. For example, imagine that several pages later in your report, you borrow from Robertson again. Would that be source [3]? No—it would be [1].

How entries in the references section are constructed may look complex. The best approach is to use the examples in Figure 6-9. They've been carefully selected to include the most common variations. Model your entries after these. In the examples, notice particularly:

- Names of books, journals, and magazines are in italics (use underscores if you don't have italics).
- Titles of articles in journals or magazines are put in double quotation marks.
- For books, list the city of publication, followed by a colon, the name of the publisher (but without all those "Co.," "Inc.," and "Ltd." words), followed by a comma, the year of publication, and a period.
- For each article, provide the date, volume number, and issue number that the article appeared in and the beginning and ending page numbers of the article. (See the style used in Figure 6-9.)

For unusual sources not illustrated here, consult *IEEE Information for Authors*.

TEXTUAL REFERENCES

Indicating the source of borrowed information in the running text of an engineering report is simple—you construct "textual references," those bracketed

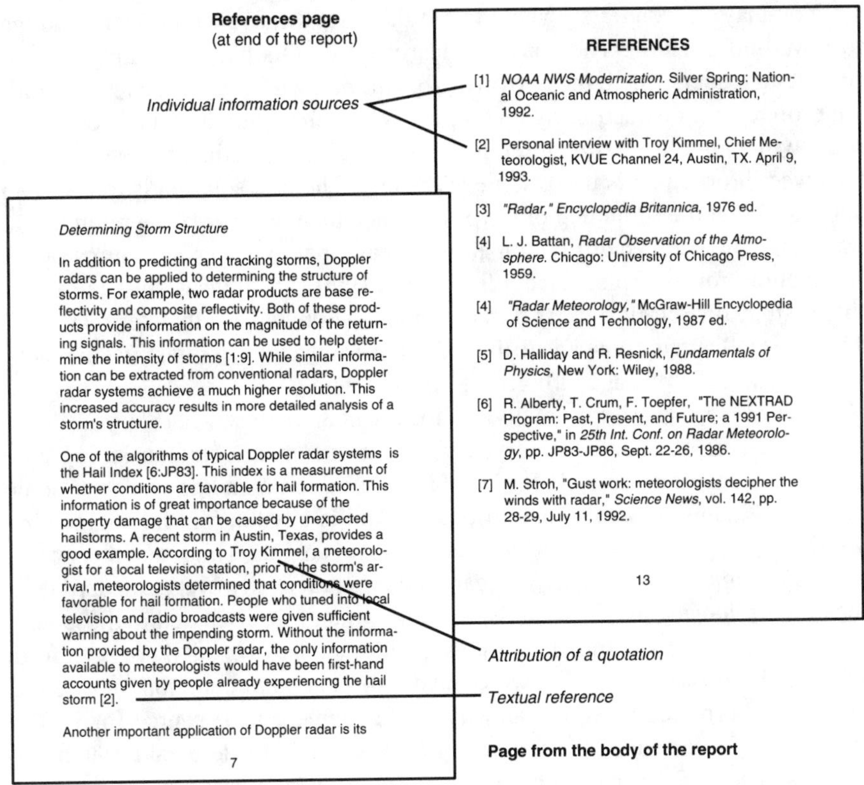

REFERENCES

[1] *NOAA NWS Modernization.* Silver Spring: National Oceanic and Atmospheric Administration, 1992.

[2] Personal interview with Troy Kimmel, Chief Meteorologist, KVUE Channel 24, Austin, TX. April 9, 1993.

[3] *"Radar," Encyclopedia Britannica*, 1976 ed.

[4] L. J. Battan, *Radar Observation of the Atmosphere.* Chicago: University of Chicago Press, 1959.

[4] *"Radar Meteorology,"* McGraw-Hill Encyclopedia of Science and Technology, 1987 ed.

[5] D. Halliday and R. Resnick, *Fundamentals of Physics*, New York: Wiley, 1988.

[6] R. Alberty, T. Crum, F. Toepfer, "The NEXTRAD Program: Past, Present, and Future; a 1991 Perspective," in *25th Int. Conf. on Radar Meteorology*, pp. JP83-JP86, Sept. 22-26, 1986.

[7] M. Stroh, "Gust work: meteorologists decipher the winds with radar," *Science News*, vol. 142, pp. 28-29, July 11, 1992.

13

Determining Storm Structure

In addition to predicting and tracking storms, Doppler radars can be applied to determining the structure of storms. For example, two radar products are base reflectivity and composite reflectivity. Both of these products provide information on the magnitude of the returning signals. This information can be used to help determine the intensity of storms [1:9]. While similar information can be extracted from conventional radars, Doppler radar systems achieve a much higher resolution. This increased accuracy results in more detailed analysis of a storm's structure.

One of the algorithms of typical Doppler radar systems is the Hail Index [6:JP83]. This index is a measurement of whether conditions are favorable for hail formation. This information is of great importance because of the property damage that can be caused by unexpected hailstorms. A recent storm in Austin, Texas, provides a good example. According to Troy Kimmel, a meteorologist for a local television station, prior to the storm's arrival, meteorologists determined that conditions were favorable for hail formation. People who tuned into local television and radio broadcasts were given sufficient warning about the impending storm. Without the information provided by the Doppler radar, the only information available to meteorologists would have been first-hand accounts given by people already experiencing the hail storm [2].

Another important application of Doppler radar is its

7

Figure 6-9 IEEE documentation system: The code numbers in the text of the report are keyed to the references page. Notice how the borrowed information is marked at the beginning by the attribution and at the end by the textual reference (the brackets).

things in the running text of a report (see the example page in the left portion of Figure 6-9).

- To indicate just the source, put the source number in brackets, for example, [3]. This tells the reader to check the references page for source 3.
- To indicate the source and the page, add a colon and the page number, for example, [3:116]. This tells the reader the borrowed information came from source 3, page 116.
- To indicate that the borrowed information came from a range of pages, add a hyphen and the end page number, for example, [3:116–122].
- To indicate that the borrowed information came from separate pages, not a range, use commas, for example, [3:116, 120, 122].
- To indicate that you've borrowed and merged borrowed information from two or more sources, use semicolons, for example, [3;7].

You may be wondering where to put the textual reference in relation to the borrowed information. There are no clear rules on this matter. Your goal is to indicate to readers where the borrowing begins and where it ends. But you don't want to create a distraction by ending every sentence with a textual reference. Some writers put the textual reference at the beginning of the passage in which borrowed information is used; some at the end. The best solution is to insert an "attribution" at the beginning of the passage and then put the textual reference in brackets at the end. An attribution is that phrase that indicates the source of the information: for example, "According to the 1990 U.S. Census Bureau . . ." or "In his study on SNMP sockets-based protocols, Edmund Smith notes that . . . ." (Figure 6-9 illustrates how an attribution phrase and a textual reference in brackets can mark off a passage of borrowed information.)

There is one last issue involving documentation: Just what do you document? The rule of thumb is that you need not document common knowledge. But what is "common knowledge"? What may be common knowledge to some may not be common knowledge to others. And is anything in the engineering world "common knowledge"? Consider several examples. Think of a theory you learned in engineering school: You can find it in practically every standard textbook on the subject, and it is not documented when it is discussed in those textbooks. That's common knowledge. But think of a controversial theory put forth by an engineer who is well known in his field. That's *not* common knowledge, and if you borrowed it, you would have to document your source for it. The same would be the case for another engineer who had made breakthrough discoveries. The difference then comes down to your familiarity with your field, whether you can distinguish common knowledge from the knowledge that is identified with specific individuals.

EXERCISES

Interview at least three professional engineers concerning the formal reports that they write, or ask to borrow examples of the reports they write. Ask the following questions or examine the example reports in the following ways:

1. How does the format of the engineering reports they create compare with the format shown in this chapter or with that specified by the American National Standards Institute's *Scientific and Technical Reports: Organization, Preparation, and Production*?
2. What are the common audiences for the reports? Are they fellow engineers or non-specialists?
3. Typically what purposes do the reports have? What functions do they perform for the engineering firm?

4. How are the graphics that are present in the reports created—by graphics specialists or by the engineers themselves?
5. How much are the reports a product of team writing—a group of engineers working on the same report together?
6. How much library research is typically required to produce the reports? How much information for the reports comes from print and nonprint sources?
7. What process does the engineering firm use in the production of reports? Do they use technical writers, graphics specialists, document designers, and editors; or is the production of reports mostly the responsibility of the engineers themselves?

BIBLIOGRAPHY

The following books and articles provide more information on the topics covered in this chapter:

American National Standards Institute. *Scientific and Technical Reports: Organization, Preparation, and Production*. ANSI Standard No. Z39.18-1987.

Beer, D. *Writing and Speaking in the Technology Professions: A Practical Guide*. New York: IEEE Press, 1991.

Buehler, M. F. "Report Construction: Tables." *IEEE Transactions on Professional Communications*. 1977: 20 (1), 29–32.

Carliner, S. "Lists: The Ultimate Organizer for Engineering Writing." *IEEE Transactions on Professional Communications*. 1987: 30 (4), 218.

IEEE Information for Authors. Piscataway, NY: IEEE Press, 1966.

Lefferts, R. *How to Prepare Charts and Graphs for Effective Reports*. New York: Barnes and Noble, 1982.

Lichty, T. *Design Principles for Desktop Publishers*. Glenview: Scott, Foresman, 1989.

MacGregor, A. J. *Graphics Simplified: How to Plan and Prepare Effective Charts, Graphs, Illustrations, and Other Visual Aids*. Toronto: University of Toronto Press, 1979.

Robertson, B. *How to Draw Charts and Diagrams*. Cincinnati: North Light Books, 1988.

7

ACCESSING ENGINEERING INFORMATION

You probably don't need to be reminded that scientific information is growing at breakneck speed—according to some estimates doubling every 2 or 3 years—while the electronic pathways to this knowledge are also rapidly expanding. The information explosion is now a constant state of affairs, and certainly a way of life for engineers. Moreover, boundary lines between science, engineering, and society are increasingly becoming blurred. Civil engineers, telecommunications engineers, and geologists may combine forces to build a very large antenna array near a city—and at the same time interact with biologists and the general public on such matters as the economic and environmental impact of their work. Even if you work in a highly specialized field like these engineers, you may need to access information from fields other than your own. To help you do that, this chapter explores the main kinds of engineering information available for your reference and research.

SOME BASIC SEARCH STRATEGIES

PREPARING FOR THE SEARCH

Although books and journals are still important sources of information (and are usually what we associate with the traditional library), they are no longer the sources we use most. Libraries themselves are currently in a transitional period as more material is put online and the worldwide information web offers more and more information through electronic access tools. Yet this doesn't mean that printed books are soon going to disappear; in fact, more printed material is being published today than ever before.

Few engineers have the leisure to browse around in a library until they stumble on the right article, book, or report. When you need information, you should first spend some time focusing on what it is you need and where it might be. Systematically ask yourself these questions:

- What is my purpose?
  - an internal report
  - a design problem
  - research
  - equipment or product selection
- What kind of information do I need?
  - practical
  - theoretical
  - economic or public policy
  - proprietary
  - product information
- What exactly do I need?
  - raw data
  - an overview of the subject
  - historical information, e.g., for product liability
  - up-to-date state-of-the-art information
  - competitive intelligence: what is our competition up to?
  - intellectual property information
  - patents
  - trademark
- What is my time frame?
  - hours
  - days
  - weeks
  - months
- What information resources do I have access to?
  - nearby experts
  - publications that colleagues and I have stacked away
  - a company library
  - electronic access (local area network, Internet, modem access)

 - a technical, college, university, or public library
 - a technical book store
- Am I willing to pay for the information by, for example:
 - buying relevant books
 - hiring a professional searcher to find what I need
 - paying for a full-text electronic search

Your answers to these questions determine where you will look for information. Remember that practically any information is available *if* you have enough time and money. Although you may prefer general information at first, try to be as specific as possible from the beginning of your search; this will save you time and money in the long run.

FOLLOWING THE TRAIL

When you need a *general background or history* on a subject, such as lasers, use the most readily available tools first. These are usually technical encyclopedias, handbooks, books, and periodicals. Be aware of the publication dates of such sources. Ask yourself if information on your topic existed prior to the publication's date. You are not likely to find much about laser surgery in anything printed before 1950; however, information on helical springs or internal combustion engines from the 1950s (or even 1930s) might be fine.

If you want *specific rather than general information,* be as precise as possible. Look for exactly what you want first; you can always become more general again later if necessary. A good technique is to figure out the hierarchy of your topic: What is more specific and what is more general. This technique works with all kinds of tools including encyclopedias, book indexes, periodical indexes, and electronic access tools. If your topic is photovoltaic cells, for example, that hierarchy could be a subdivision of any of the following, depending on your focus:

1. **Solar energy**—if you're working on renewable energy sources
2. **Thin film technology**—when you need information on how the cell is manufactured
3. **Marketing**—if you want to know how many cells are sold each year
4. **U.S. energy grid**—if you are interested in power systems

In a nutshell, how and where you look depends on your purpose or task. Remember, however, that although libraries and electronic information sources are good places for general, historical, theoretical, or scholarly technical information, they are poor places to look for proprietary information.

If you are in a library, don't be afraid to ask for help. Many libraries have staff who are experts in carrying out an information search and who are most willing to help you when courteously approached. Engineering librarians often suggest that users apply the *20-minute rule:* If after looking for information for 20 minutes you find nothing relevant, ask for reference help (in most libraries this means a trained librarian, not a clerk).

Become proficient with the gopher "search engines" on the Internet, such as Archie and Veronica, and those on the World Wide Web, such as Lycos or WebCrawler. An amazing amount of engineering data is now available from these sources, and more becomes available every day. You are likely to need skilled help to access much of this information at first, and there may be a charge for using some databases. (See "Internet Engineering Information Resources" for some starting points.)

SOURCES OF ENGINEERING INFORMATION

GENERAL BOOKS

In the United States alone more than 50,000 book titles are now published annually, compared to 20,000 in 1960. Although all the information contained in these hardcopy publications could theoretically be reproduced or replaced by information in electronic, machine-readable form, this is not likely to happen for a good many years. You can still expect to find plenty of worthwhile information in library stacks and periodical rooms.

When you are researching a particular topic, books can provide excellent background information. A quick look through a book's table of contents will give you a good idea of whether it's likely to have what you are looking for, and you can fine-tune the search by studying the book's index.

Obviously, the most recently published books are going to give you the best picture of a current area of technology, but some older books may provide excellent background to a field. For example, a book or encyclopedia published in the 1960s on radio wave propagation and the ionosphere might still contain some useful background information. For many current research topics, however, books tend to be too general. To obtain more specific information on technological advancements, you must go to journal articles, technical reports, or other sources described later in this chapter.

As you probably know, many libraries have moved their catalogs from the traditional physical card catalog to online media. Some libraries have gone a step further and made their online catalogs accessible remotely—in many cases, through the Internet. This means that if you cannot find a specific book or the right books in your local library, you can check other libraries on the Internet. (What if you find something in a library 400 miles away? Ask your librarian about interlibrary loan. If waiting for the book to arrive by mail is not an option, at least you know what is available on your topic.) Online library catalogs offer you more powerful search tools than do traditional card catalogs. If you are searching for books on a topic, you are limited by what a particular librarian believed was related to that topic. On the Internet, you can sample the cataloguing ideas of many different librarians. For lists of Internet-accessible libraries worldwide, check these resources:

- Get the file **internet.library** in the directory *library* by anonymous FTP from **ariel.unm.edu**. This is a listing compiled and maintained by Ron Larsen and Art St. George.
- Get the files **libraries.intro**, **libraries.africa**, **libraries.americas**, **libraries.asia**, **libraries.australia**, **libraries.europe**, and **instructions** in the directory *pub/staff/billy/libguide* by anonymous FTP from **ftp.utdallas.edu**. These listings are compiled and maintained by Billy Barron.
- To access the Library of Congress catalog, telnet to **locis.loc.gov,** or use gopher access to **marvel.loc.gov**.

REFERENCE BOOKS

In addition to books located in the stacks, most engineering libraries (and technical information centers) have a reference section where you can find encyclopedias, handbooks, reference manuals, and similar materials. These volumes normally cannot be checked out of the building and are used for quick, on-the-spot "look-up" of factual information. The following lists just a sampling of the wide range of reference books that exist for engineers.

Kirk-Othmer's Encyclopedia of Chemical Technology, 6th ed. 1992. 25 volumes. This publication's title is somewhat misleading since the volumes cover all areas of technology. You may find especially useful the extensive references at the end of each article to patents, conference proceedings, and journal articles. Also available electronically on CD-ROM and remote access from Dialog Information Services (file 302).

McGraw-Hill Encyclopedia of Science and Technology, 7th ed., 1992. 20 volumes. Contains almost 8000 articles on science, engineering, and other technical subjects. Clear and readable, with plenty of illustrations. An excellent source to check first for general background information.

Van Nostrand's Scientific Encyclopedia, 7th ed., 1989. Concentrates on the basic and applied sciences, with over 17,000 entries arranged alphabetically in two volumes. Functions as a technical dictionary.

McGraw-Hill Dictionary of Scientific and Technical Terms, 4th ed., 1989. Provides almost 100,000 definitions of terms and includes some 2800 illustrations.

The Wiley Encyclopedia Series in Environmental Science: Energy and the Environment, 1995. Four volumes of alphabetical entries on energy-related topics relating to technology and its impact on the environment.

Handbook of Industrial Engineering, 2nd ed., 1991. Almost 3000 pages of detailed information on such topics as performance measurement, quality control, engineering economy, and manufacturing engineering.

Chemical Engineers' Handbook, 6th ed., 1984. Includes material from general mathematics and tables to specialized treatment of topics such as psychometry, process machinery, and distillation. Generally known as *Perry's.* This is a standard for petroleum and chemical engineers.

Handbook of Mechanical Engineering, 1994. Like its cousin, *Mark's Standard Handbook of Mechanical Engineering,* this handbook contains useful entries, tables, and data on all aspects of mechanical engineering and other subjects of use to mechanical engineers.

Standard Handbook for Civil Engineers, 3rd ed., 1983. Covers construction, structural theory and design, materials, and management for the various fields of civil engineering, including environmental concerns.

Standard Handbook for Electrical Engineers, 12th ed., 1987. Substantial coverage of all aspects of electrical engineering, with numerous tables, charts, and graphs.

You can also count on there being a specialized encyclopedia or handbook for just about every specialty you can imagine. Here are just four examples:

Concrete Construction Handbook, 3rd ed., 1993.

Handbook of Noise Control, 3rd ed., 1991.

Radar Handbook, 2nd ed., 1990.

Reference Manual for Telecommunications Engineering, 2nd ed., 1994.

How to Find Reference Books

The best way to find useful books and reference tools in your field is through onsite or remote electronic access to library card catalogs. For remote access to specific libraries, call the reference desk: Most will provide dial-up or Internet instructions over the phone. You can also find a library's Internet address by using one of the various "gopher" search engines such as Archie, Veronica, or graphical interface search tools such as Lycos or Webcrawler. Subscribers to America Online and CompuServe also can get easy access to electronic library catalogs. This is the simplest and most efficient way to see what books have been published on specific topics and what nearby libraries have available.

Other than using a library's electronic catalog to find useful reference tools in your field, you can go to the hardcopy version of *Sheehy's Guide to Reference Books,* which lists numerous categorized sources of information and is updated regularly. *Sheehy's* is divided into main groupings such as pure and applied sciences, social sciences, and humanities, each of which is subdivided into more specific areas such as petroleum engineering, computer science, or insurance. Some projects might lead you to several sections of *Sheehy's:* For example, you could do preliminary research on solar-heating devices by looking for titles under air conditioning and heating, electrical engineering, and energy resources. This reference tool is available in most libraries.

JOURNALS

You probably already subscribe to one or two professional journals and may have access to others through friends' subscriptions or through a nearby library. Over 10,000 paper and 200 electronic scientific and technical journals are published every year and both numbers are growing. These journals are essential for any engineer who wants to keep up with the latest developments in a given field.

The information in journals (unlike books) generally consists of highly technical short papers and reports on the most current research and thinking in an area of specialization. Few libraries can subscribe to more than a fraction of the periodicals published, and chances are that any library you use will only have a cross-section of the most respected journals, or may concentrate only on journals for a certain field. These limitations can be overcome by interlibrary loans, of course, if the journal isn't available electronically, or by commercial document delivery. Document delivery companies are in the business of providing copies of articles, usually by FAX or express delivery, for a fee. For phone numbers of relevant companies call any local library or look in the Yellow Pages under "document delivery."

To become familiar with all the journals published in your field consult *Ulrich's International Periodicals Directory,* which annually lists journal titles in some 200 categories, including engineering, which itself is further subdivided by fields such as civil, electrical, mechanical, and petroleum. Most public and college or university libraries own *Ulrich's*. It is also available from Dialog Information Services for a fee.

There are also some electronic options for journal research, such as CARL and WORLDCAT. Both are available over the Internet, and many libraries provide free local access. CARL is particularly powerful because it consists of a searchable list of current journals (1986–present) and their tables of contents, and because it permits you to order copies of articles for a fee. WORLDCAT, from OCLC, contains the titles of all books and journals owned by participating libraries.

INDEXES AND ABSTRACTS

You couldn't begin to read all the journal articles relevant to your work, yet if you want to stay aware of what's being done in your field, or need to research a particular topic, you need to know what is being written. This is why indexing and abstracting services exist. An *index* lists the subjects covered in a periodical or group of periodicals and gives the title, volume, date, and author(s) of each article published on that subject in a particular period of time. Some indexes add summaries of each article, known as *abstracts*.

Even though the example in Figure 7-1 is from the electronic version of *Engineering Index,* it is representative of the kind of information found in most paper or electronic abstracts. Coverage varies; not all articles are abstracted. You can check coverage in the preface or by selecting the Journals field on the electronic abstract.

Printed abstracts and indexes have existed for over 150 years and are still widely used. Since 1970 they have also been available in computer-stored databases, and since 1984 on CD-ROMs. Both storage mediums enable you to search for information with all the ease and speed that a keyboard and monitor permit. The advantage of using electronic abstract and indexing services is that you can enter **keywords** for the topic you are researching and call up references containing those keywords. With printed versions, you have to first find the subject area you are interested in and then look at all the listings within that area until you find relevant citations.

Indexing and abstracting services, whether in paper or electronic, are indispensable when you first research a topic. Hundreds of such services exist on practically any topic you can imagine, and most services issue updates at least once a month to keep current. You may find the *Engineering Index* the most useful, but others applicable to engineering include the following:

Title of Paper Index	**Electronic Equivalent**
Applied Science and Technology Index	ASTI
Business Information	ABI/Inform

Chemical Abstracts	CA
Computer and Control Abstracts	INSPEC
Electrical and Electronics Abstracts	INSPEC
Engineering Index	Compendex
International Aerospace Abstracts	Aerospace Database
Metals Abstracts	Metadex
Nuclear Science Abstracts	Not available.
Pollution Abstracts	Same name.

Most indexes and abstracts list articles from a large assortment of journals. For instance, the first title above, *Applied Science and Technology Index,* lists

```
(c) 1995 Engineering Info. Inc. All rts. reserv.
04141502 E.I. No: EIP95042678800
   DIALOG(R)File 8:Ei Compendex*Plus(TM)
   Title: Trajectory prediction for moving objects using artificial neural
networks
   Author: Payeur, Pierre; Le-Huy, Hoang; Gosselin, Clement M.
   Corporate Source: Laval Univ, Ste-Foy, Que, Can
   Source: IEEE Transactions on Industrial Electronics v 42 n 2 Apr 1995.
p.147-157
   Publication Year: 1995
   CODEN: ITIED6 ISSN: 0278-0046
   Language: English
   Document Type: JA; (Journal Article)
   Treatment: A; (Applications); T; (Theoretical)
   Journal Announcement: 9506W3
   Abstract: A method to predict the trajectory of moving objects in a
robotic environment in real-time is proposed and evaluated. The position,
velocity, and acceleration of the object are estimated by several neural
networks using the six most recent measurements of the object coordinates
as inputs. The architecture of the neural nets and the training algorithm
are presented and discussed. Simulation results obtained for both 2D and 3D
cases are presented to illustrate the performance of the prediction
algorithm. Real-time implementation of the neural networks is considered.
Finally, the potential of the proposed trajectory prediction method in
various applications is discussed. (Author abstract) 11 Refs.
   Descriptors: *Neural networks; Robotics; Real time systems; Learning
algorithms; Learning systems; Computer simulation; Three dimensional;
Performance; Manipulators; Mathematical models
   Identifiers: Moving objects; Trajectory prediction; Object coordinates;
Two dimensional
   Classification Codes:
   723.4 (Artificial Intelligence); 731.5 (Robotics); 722.4 (Digital
Computers & Systems); 723.1 (Computer Programming); 723.5 (Computer
Applications); 921.6 (Numerical Methods)
   723 (Computer Software); 731 (Automatic Control Principles); 722
(Computer Hardware); 921 (Applied Mathematics)
   72 (COMPUTERS & DATA PROCESSING); 73 (CONTROL ENGINEERING); 92
(ENGINEERING MATHEMATICS)
```

Figure 7-1 Example of the abstract of an article listed in *Engineering Index.*

references from journals including *Industrial Wastes, Industrial Robots, Journal of Engineering for Power, Journal of Metals, Food Technology, Computer Design, Automotive Engineering, Robotics Age, Welding Journal, Datamation,* and *Nuclear and Chemical Waste Management.*

TECHNICAL REPORTS

Hundreds of thousands of technical reports are written each year in the United States alone; many are available on electronic media or on microfiche. A technical report may be similar to a paper presented at a conference or to a journal article, but may be a lot longer. Technical reports are usually written by specialists for other specialists and report on the results of research and development; in most cases, they are highly technical. Reports sponsored by a government grant or contract are the easiest to find, whereas proprietary and classified reports are not generally available.

Because so many reports constantly flow into the ocean of technical information, you must go to indexes and abstracts to narrow your search for reports in a certain field or for a specific report. This is one area where paper indexes are being replaced by electronic indexes such as NTIS and NASA RECON, described below.

NTIS (National Technical Information Service). This electronic or CD-ROM indexing service is updated twice a month, and adds thousands of new reports of completed government-sponsored research to its database each year from organizations such as NASA, the Department of Energy (DOE), and the Environmental Protection Agency (EPA). NTIS is the major source for information on nonproprietary and unclassified reports sponsored by government agencies and contractors. It lists the subject of each report, its individual and corporate author, and the contract and report number.

See Figure 7-2 for an illustration of a typical NTIS record. Notice that the paper described is part of an internal Langley Research Center report and that the entire report must be purchased (see **order as**). Reports labeled with a **PC** (or price code) can be ordered from NTIS (1-800-336-4700) in paper or microfiche. Price-code specifics are available from NTIS or any large library. Unfortunately many libraries buy NTIS reports very sparingly; they are rarely available through interlibrary loan.

NASA RECON. This electronic index is part of NASA's Information System and contains indexes of technical reports written on research done for NASA and released by government agencies, U.S. and foreign institutions, private companies, and universities. It also covers patents, dissertations, and translations and provides an index by subject, individual and corporate author name,

```
1824205 NTIS Accession Number: N95-16475/2/XAB
   Simulation of the Coupled Multi-Spacecraft Control Testbed at the
Marshall Space Flight Center
   Ghosh, D. ; Montgomery, R. C.
   National Aeronautics and Space Administration, Hampton, VA. Langley
   Research Center.
   Corp. Source Codes: 019041001; ND210491
   Oct 94 22p
   Languages: English
   The Role of Computers in Research and Development at Langley
   Research Center p. 497-517.
   NTIS Prices: (Order as N95-16453/9, PC A99/MF A06
   Country of Publication: United States
   The capture and berthing of a controlled spacecraft using a robotic
manipulator is an important technology for future space missions and is
presently being considered as a backup option for direct docking of the
Space Shuttle to the Space Station during assembly missions. The dynamics
and control of spacecraft configurations that are manipulator-coupled with
each spacecraft having independent attitude control systems is not well
understood and NASA is actively involved in both analytic research on this
three dimensional control problem for manipulator coupled active spacecraft
and experimental research using a two dimensional ground based facility at
the Marshall Space Flight Center (MSFC). This paper first describes the
MSFC testbed and then describes a two link arm simulator that has been
developed to facilitate control theory development and test planning. The
motion of the arms and the payload is controlled by motors located at the
shoulder, elbow, and wrist.
   Descriptors: *Attitude control; *Computerized simulation; *Control theory;
*Dynamic control; *Manipulators; *Robot arms; *Space shuttles; *Space
stations; *Spacecraft configurations; *Spacecraft control; *Spacecraft
docking; Equations of motion; Ground tests; Payloads; Robotics; Shoulders;
Space missions; Wrist
   Identifiers: NTISNASA
   Section Headings: 84A (Space Technology--Astronautics)
```

Figure 7-2 A typical record available from NTIS (National Technical Information Service).

and contract and report number. Some of the reports are classified or proprietary, but most are available in large university engineering libraries or may be purchased from NTIS itself. Some are available on the Internet for free. Many university libraries provide onsite access to the entire database for a small fee.

PATENTS

For most engineers, patent documents are a rich source of technical and scientific information. Patents describe in detail the designs, materials, machines, and processes of new and useful inventions, and in return the inventor is granted a right of ownership by the government that limits others from making, using, or selling the patented item throughout the United States for 20 years (patents issued prior to June 1995 are good for only 17 years). The U.S. Patent Office granted three patents in 1790, the year it was created; the number of U.S. patents now granted annually is close to 150,000. As of 1995, some 5.5 million patents have been issued.

You might want to do a patent search to

- find out about a specific patent
- learn about recent inventions in a particular field
- find out if your invention has already been patented
- gain ideas for further development of your invention
- see what inventions known competitors have patented

Many engineers are unaware of the enormous amount of technical information contained in patent documents. In fact, it's estimated that most of the technology described in U.S. patent information is found in no other source. For this reason, learning to perform patent searches is well worth your effort. Even though it may be a time-consuming and arduous process, the wealth of information you can find more than compensates. Because patent searching is complex, read about the process in one of these two general sources:

General Information Concerning Patents: A Brief Introduction to Patent Matters. U.S. Department of Commerce, Patent and Trademark Office: Washington, D.C. 1992.

Timothy Wheery's *Patent Searching for Librarians and Inventors,* ALA, 1995. This publication demonstrates the important differences between copyrights, patents, and trademarks.

The best places to find recent patent information are the *Official Gazette (OG) of the United States Patent and Trademark Office (USPTO),* the USPTO home page, and commercial databases such as Lexis/Nexis, or U.S. Patents Fulltext. The *Official Gazette (OG)* contains brief descriptions and drawings of the 1500 patents granted every Tuesday. A complete electronic copy of patents issued since 1994 can be found on the Internet at the following web site: **http://town.hall.org/patent/patent/patent.html**.

Because the data is so voluminous, an annual paper index to the *Official Gazette* comes out in two volumes entitled *Index of Patents.* It is available at many large public and university libraries. A more efficient way to search the patent literature is to go to a Patent and Trademark Depository library (PTDL). A list of these libraries can be found in the *OG* or on the USPTO's home page (**http://www.uspto.gov/**). Other information can also be found on this home page. PTDLs provide free access to all tools including the *OG*, *CASSIS* (a computerized index), and copies of issued U.S. patents.

If you've never seen a U.S. patent, look at Figure 7-3 and notice how much information you can find on just the front page. Each front page includes the inventor's name (patentee), owner at date of issuance (assignee), date issued (needed to compute expiration), citations to other relevant patents and articles, one drawing, and an abstract. Besides the front page, a patent (which averages ten pages) has two other sections. In the disclosure section, the inventor

US005392125A

United States Patent [19]

Reisser

[11] **Patent Number: 5,392,125**

[45] **Date of Patent: Feb. 21, 1995**

[54] **INSTRUMENT FOR DETERMINING VISUAL SURFACE PROPERTIES**

[76] Inventor: **Helmut Reisser,** Rosenstrasse 6, D-8026 Ebenhausen, Germany

[21] Appl. No.: **981,327**

[22] Filed: **Nov. 24, 1992**

[30] **Foreign Application Priority Data**

Nov. 25, 1991 [DE] Germany 4138679

[51] **Int. Cl.**[6] **G01N 21/55**

[52] **U.S. Cl.** **356/445;** 356/446

[58] **Field of Search** 356/446, 448, 445, 447; 250/227.29

[56] **References Cited**

U.S. PATENT DOCUMENTS

3,012,465 12/1961 Goldberg 356/136
3,999,864 12/1976 Nutter 356/212
4,285,597 8/1981 Lamprecht et al. 356/446
4,988,205 1/1991 Snail 356/446

FOREIGN PATENT DOCUMENTS

0020971 1/1981 European Pat. Off. .
0095759 12/1983 European Pat. Off. .
356935 6/1980 Germany .
1190564 5/1970 United Kingdom .
1404573 9/1975 United Kingdom .
2189881A 11/1987 United Kingdom .
2192454A 1/1988 United Kingdom .
2242977A 10/1991 United Kingdom .

OTHER PUBLICATIONS

Bruecker, "Measuring The Specular Reflection on Surfaces", *International Laboratories,* pp. 28–32, (1990).

Primary Examiner—Richard A. Rosenberger
Assistant Examiner—Robert Kim
Attorney, Agent, or Firm—Foley & Lardner

[57] **ABSTRACT**

An instrument for determining visual surface properties includes a generally cylindrical housing **10** with an illumination unit **12** and a photo cell **22** disposed in its upper portion. The beam A emitted by the illumination unit **12** passes through an optical system **19** disposed in a lower portion of the housing **10** and is directed by a first prism **20** at a predetermined angle onto the surface **11** under examination. The beam A' reflected at a predetermined angle by the surface **11** is deflected by a second prism **20'**, passes the same optical system **19** a second time and is received by the photo cell **22**. A plurality of pairs of prisms with different angles of incidence and reflection are arranged concentrically about the central axis N of the housing **10**. An inexpensive instrument is thus achieved which is small in the direction parallel to the surface under examination.

7 Claims, 2 Drawing Sheets

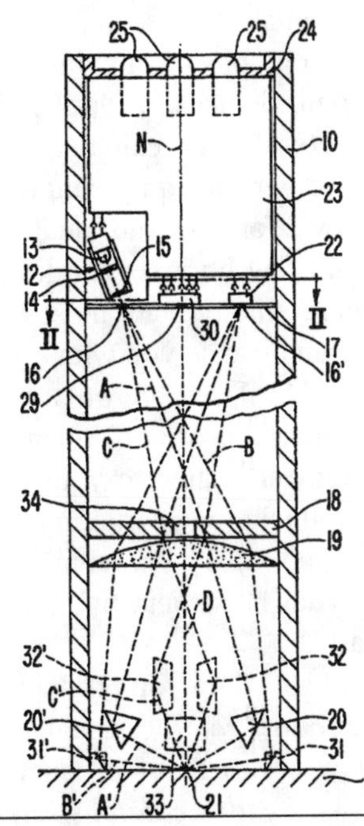

Figure 7-3 The Front Page of a U.S. Patent Document.

Source: United States Patent and Trademark Office (Washington, D.C., February 21, 1995).

describes or "discloses" how his or her invention works and how it relates or improves on existing solutions to the same problem. The final and most important part is the claims section. Here the inventor gives the legal description of what is actually protected by the patent.

For information on patents issued in other countries, the best sources are online. The cost varies widely. Three useful sources are

JAPIO for Japanese patents

INPADOC for European patents

DERWENT for world patents

If you are interested in applying for a patent for your own work, two good sources of information to begin with are

Susan Ardis, *Introduction to U.S. Patent Searching.* Libraries Unlimited, 1990.

David Pressman, *Patent It Yourself.* Nolo Press, 1995. (Text and software are available.)

PRODUCT LITERATURE

A gold mine of information for engineers can be found in product literature, which includes product, manufacturer (company), and vendor catalogs, as well as product selectors, buyers' guides, and so on. The information in these may include performance data, photographs or drawings of products, data books for computers and integrated circuit devices, or application notes and other information about specific products. Topics can range from aerospace ordnance equipment to transportation and vehicle equipment or supplies. Product material differs tremendously in shape, format, and quality. A lot of it constantly changes, much is undated, and few libraries can keep up with all the material available. Of course, sales representatives are most willing to help you get what you want, and most libraries can provide you with company addresses.

This material is indispensable if you are on a design project. For instance, you may want to know the dimensions or performance figures for specific components, accessories, or equipment related to your project. You will also need this information if you are producing or marketing your own company's products. You can save time and money by knowing what is already available in your field that precisely fits your needs and by being able to compare various products currently on the market.

Product literature is aimed at selling you products and only briefly provides details of specific products. *Product catalogs* usually feature one product, while *manufacturers' catalogs* show a variety of products for sale from a specific manufacturer (see Figure 7-4). *Vendor catalogs,* on the other hand, show all

3

Custom, Modified, and Special Valves

For the customer with highly individual requirements, Huntington will custom design and fabricate special purpose valves for specific applications. Original equipment manufacturers are invited to submit specifications for their particular valve requirements.

Various modifications to standard valves are available when specified.

Copper Bonnet Seals. Viton Valves with OFHC copper bonnet seals are often requested where toxic, corrosive, or radioactive materials are involved.

Mechanical Position Indicators. Limit Switch mechanical position indicators are available as options on all Huntington pneumatic valves.

Flange Options. Valve ports may be fitted with any type of flange, no flange, or any combination of these options.

Adapter Valves. Butterfly valves may be modified to perform an adapter function such as the valve pictured below which has a KF Style fitting on one side and a hose fitting on the other. (Used on mechanical vacuum pumps).

Special Ports. Roughing ports, thermocouple ports, or gas backfill ports can be added to any valve.

Special straight through valve with roughing port.

The High Temperature Kalrez Option. Kalrez seals retain their elastic properties at temperatures approximately 100° C higher than the limits of other elastomers-- as high as 288° C in long term service and up to 316° C with intermittent use. Thermally stable at lower temperatures also, Kalrez is particularly suitable for applications requiring cycling between low and high temperatures.

Compared with metal seals, Kalrez O-rings are easily installed and conform to any sealing surface irregularities that may occur due to reassembly or wear.

Compared with O-rings made from other elastomers including Viton, to which it is most similar, Kalrez O-rings are more resistant to swelling and embrittlement and retain their elastomeric properties longer, especially at higher temperatures.

Kalrez perflouroelastomer is an amorphous, low modulus rubber combining the elastomeric qualities of Viton with the chemical resistance of Teflon. It works effectively in difficult environments where other sealing materials fail. It resists attack by nearly all chemical reagents, including ethers, solvents, ketones, esters, amines, oxidizers, fuels, acids, and alkalis, and thus performs long term service in certain process streams where other elastomers would be quickly destroyed.

Huntington will also supply Teflon, Silicon, Buna-N, Neoprene, or other available sealing material upon special request.

Exhaust Mufflers. Exhaust from pneumatic valves can often be distracting in the laboratory. Any Huntington pneumatic valve may be equipped with an exhaust muffler that easily screws into the solenoid.

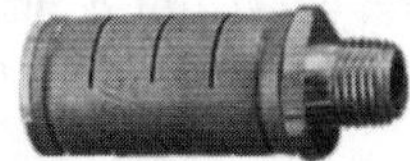

Convenient Valve Wrench. For your convenience, the Huntington valve wrench, Model VW-15, (priced at $12.00), facilitates removal of the threaded actuator on manual or pneumatic elastomer-sealed valves.

☎ (415) 964-3323 • FAX (415) 964-6153 **HUNTINGTON LABORATORIES • 67**

Figure 7-4 A page from Huntington Laboratories' catalog, Huntington Mechanical Laboratories, Inc. 1994 (reprinted by permission).

products for sale by the vendor and are designed for fast and easy comparison between several competing brands. Any of these usually provide purchasing information such as the manufacturer's location and phone number, and most give at least limited product specifications, performance data, drawings, test data, and application details.

One example of a product catalog dedicated to a specific field is the *Electronic Engineer's Master* catalog (EEM). Designed to help users see a range of similar products, EEM is a collection of pages from catalogs of companies around the world. It provides an advertiser's index and a manufacturer's index, including local sales offices and distributors.

To get an idea of the enormous variety of products described in catalogs, look at the annual *Thomas Register of American Manufacturers,* commonly called the *ThomCat.* The first sixteen volumes allow you to access the names, addresses, and telephone numbers of about 150,000 U.S. manufacturing firms. You can look up companies that make a specific product—for example, backhoes. Other volumes of interest include a U.S. tradename index; you can use this index to find out who manufactures Teflon, for example. Volumes 19–26 consist of selected pages from individual company catalogs. Each section of the *ThomCat* is fully cross-referenced with other sections. You can get copies of most catalogs described in the *ThomCat* directly from the publisher.

Other useful sources of product information are the large microfilmed collections of manufacturers' and industrial catalogs such as *Information Handling Service's Visual Search Microfilm Files,* and *Information Marketing International.* These consist of copies of complete catalogs along with indexing, permitting you to look up several manufacturers of potentiometric multimeters, for example.

STANDARDS AND SPECIFICATIONS

Most products we use daily are designed and produced in accordance with certain standards or specifications. The length of toothpicks, the softness of toilet paper, and the different grades of sandpaper are all controlled by agreed-upon industrial standards. These standards are essential if you want to be able to consistently fit light bulbs into sockets all over the country, screw nuts on to bolts, replace engine parts, or rely on the strength of concrete. Furthermore, we live in a society that is both safety conscious and increasingly alert to the quality of mass-produced consumer items. Thus, the setting of industrial and other standards, and the incorporation of them into the work of engineering, can only increase—whether imposed by consumers, industry, or government.

Whenever you design something as an engineer, you must be aware of what standards, specifications, or codes already exist that might be relevant to your product. One professional engineer puts it this way:

You will usually be informed of the applicable specs by your managers, but they may miss some. If you do not comply with an applicable spec, you will have to redesign, with cost to your organization and criticism of yourself whether or not it was your fault. It is good practice to assure yourself independently that you know all applicable specs.

Source: Lawrence L. Kamm, *Successful Engineering: A Guide to Achieving Your Career Goals.* New York: McGraw-Hill, 1989, 145.

The terms *standards* and *specifications* are often used as synonyms because they can both refer to the guidelines by which something is measured, designed, tested, or manufactured. When a formal distinction is made between the two terms, however, "standard" is more general while "specification" is more specific. You might specify, for example, exactly how you want your house built, but the builder will still have to stay within the broader construction standards set by your neighborhood association, city, state, or other entities.

The standards for specific products are set by the trade associations, companies, manufacturers and professional societies involved in those products, and also by government agencies and international standards organizations. Among the hundreds of organizations that produce standards are the following:

American Gear Manufacturers Association (AGMA)
American Society of Mechanical Engineers (ASME)
American Society for Testing and Materials (ASTM)
Institute of Electrical and Electronics Engineers (IEEE)
National Wire Rope Manufacturers Association (NRMA)
Underwriters Laboratory (UL)
American National Standards Institute (ANSI)—U.S. treaty representative to other standards producers

The main U.S. government generators of standards are the General Services Administration (GSA) and the Department of Defense (DOD). Two of the major non-U.S. organizations that collect and formulate engineering standards are the German Institute for Standardization (DIN) and the International Organization for Standardization (ISO) in Geneva, Switzerland.

So many engineering standards exist, and so many different organizations issue them, that you might think finding a standard is like looking for the eye of a needle in a cosmic haystack. Fortunately, this is not the case. Nowadays, engineers have access to efficient ways to locate standards information. You can go to traditional printed sources for each organization, like the *Catalog of American National Standards (ANSI)* or *Index of Specifications and Standards (DOD).* Another way is to use a more general tool which covers several organizations, such as *Industry Standards and Engineering Data: Number and Subject Index,* from Information Handling Service (IHS). You can also access standards on the Internet: standards for the International Telecommunications Union can be found at **http://www.itu.ch**; standards for the International Organization for Standardization, at **http://www.iso.ch**; and standards for the American National Standards Institute, at **http://www.ansi.org**.

Figure 7-5 shows the typical structure of a standard. Notice in this brief example that "resolution" has a number of subdivisions and that four different organizations are involved. This illustration also demonstrates the very systemized nature of standards numbers. For example, ASTM F807-83 consists of first the organization (ASTM), then the standard number, (F807), and a hyphen followed by the year issued (83). And if you were concerned about the validity of this particular standard, you'd pay attention to the parenthetical "R 1993." This indicates that the issuing organization reviewed and reapproved this 1983 standard in 1993. You should always check to be sure you are using the most up-to-date standard, unless of course you are involved with a historical design problem and need the standard that was current at the time the equipment was designed (or failed).

Other electronic resources for standards research also exist. You can use electronic indexes to search for standards by subject, number, or organizational name. These kinds of indexes are especially useful if you are looking for the standards of a particular organization or need to perform a general search. Two of the companies providing such electronic standards research are *Information Handling Services* (which covers industry and international standards as well as military and federal ones), and the *National Standards Association,* which

```
Resolution:
Used for:  Limiting Spatial Resolution: LSR; Resolving Power
See also:  Image Intensifiers: Numerical Aperture

Tolerances
   ANSI PH3>609-80 Dimensions for Resolution Test Target for Photographic Optics
   (R 1987) NFP(A) T2.9.6 R1-90. Hydraulic Fluid Power--Calibration Method for
   Liquid Automatic Particle Counters. (Revision/Re designation ANSI
   B93.28-1973)

Cathode-Ray Tubes
   EIA TEB25-85 Survey of Data-Display CRT Resolution Measurement Techniques.
Copying Machines
   ASTM F807-83 Standard Practice for Determining Resolution Capability of
   Office Copiers. (R 1993)
Image Processing Systems
   AIIM TR26-93 Resolution as it Relates to Photographic and Electronic Imaging.
Photographic Lenses
   ANSI PH3.63-74. Method for Determining the Photographic Resolving Power of
   Photographic Lenses.  (R 1991)
```

Figure 7-5 Description of an industrial standard from *Industry Standards and Engineering Data: Subject Index.*

focuses on government and industry standards and specifications. The National Standards Association's *Standards and Specifications* also provides an electronic reference collection of almost a quarter of a million standards and specifications (including FEDSpecs and MILSpecs) and for a fee will provide indexes based on a keyword search.

U.S. GOVERNMENT SPECIFICATIONS

Since the U.S. government is one of the world's largest buyers of practically every kind of civilian and military product or service, it has produced enormous quantities of standards and specifications describing the requirements for its purchases. In 1995 a policy decision was made concerning FEDSpecs and MILSpecs. The U.S. Congress and the GAO instructed government agencies to use ASME, ASTM, ANSI, IEEE, UL, and other standards or specifications whenever possible. If you are interested in the specifications for items bought by the General Services Administration and the U.S. armed forces, you should begin with the following:

United States Federal Standards and Specifications (FEDSPECS)
United States Military Standards and Specifications (MILSpecs)
Department of Defense Index of Specifications and Standards (DODISS).

Some engineering libraries have an Industry Standards Collection containing information on the standards to be used by the government that are formulated by organizations like IEEE, ANSI, and ASME (American Society of Mechanical Engineers). Those with large collections may also provide you with access to one of the electronic indexes mentioned above. Sometimes you can get help directly from the issuing agency, especially if you are buying a copy of the document. Commercial vendors, such as Global (1-800 624-3974), provide express-service copies of U.S. and international standards to those who need them.

INTERNET ENGINEERING INFORMATION RESOURCES

To this point, we have discussed mostly print-based methods of finding information, some of which have electronic analogs, such as *Engineering Index* on CD-ROM. However, the Internet now offers some powerful alternatives for information research to the engineer.

USENET NEWSGROUPS

Newsgroups on the Internet are ongoing discussions focused on a particular topic or field and used by people all over the world. For example, as a civil engineer, you might be interested in the discussion on the **sci.engr.civil** newsgroup. (For a sampling of engineering-related newsgroups, see Ch. 4.) Newsgroup activity (a series of e-mail messages all strung together) is generally not saved on the host computer; using what exists, however, you can do a search for topics you are interested in. Of course, you must be radically skeptical of any information you find in a newsgroup. Like any other e-mail, it is written rapidly and with little careful revision; and you cannot be sure of the qualifications of the senders. Still, newsgroups are a good way to get some information and opinions and, more importantly, some names and addresses to contact. You can send e-mail to these groups asking for information, opinions, or contacts on your topic.

How do you find out what newsgroups exist for your field or topic? USENET newsgroups will never be anything but chaotic. A perfect newsgroup may exist on the other side of the country; but if your local newsfeed doesn't carry it, you may never hear of it. All you can do is work your way through the ones that your local newsfeed carries (which is determined by the system administrator of that computer), looking for and then subscribing to the ones that are of interest. There are, however, ongoing attempts to maintain directories of active USENET newsgroups:

- Check out the USENET newsgroups: **news.groups, news.announce.newusers, news. groups**.
- Also, you can try getting the file **Alt_Newsgroup_Hierarchies** in the directory *pub/usenet/news.announce.newusers* by anonymous FTP from **pit-manager.mit.edu**.

ELECTRONIC MAILING LISTS

Mailing lists are another way to do some research on the Internet. For example, there is a list called CIVIL-L for civil engineers. Here's how it works: When you subscribe to it, you receive any e-mail that anyone else subscribed to CIVIL-L sends to that list; and any e-mail you send to the list gets "reflected" to all the other subscribers to CIVIL-L. Many, but not all, of these electronic mailing lists "archive" their e-mail activity. Not only can you watch current e-mail for your topic, you can search these archives for your topic and see what subscribers have said about it.

As with USENET newsgroups, you need to maintain a healthy skepticism about the information you find in electronic mailing lists. Because it takes more effort to subscribe to a list than it does to engage in newsgroup activity, postings on a mailing list are likely to be more considered and professional. But still it's e-mail and it's the Internet—messages tend to be written and sent hastily, and you cannot be sure of the qualifications of the senders. Even so, the contacts you can get from following a mailing list or rummaging in its archives can be invaluable, even if you don't trust the information itself.

How do you find mailing lists in your areas of interest? Try these possibilities:

- Get the file **ACADLIST README** in the directory *library* by anonymous FTP from **ksuvxa.kent.edu**. This is a listing maintained by Diane K. Kovacs.
- Send e-mail to **listserv@bitnic.bitnet**, leave the subject line of the message blank, and in the body of the message write **list global**.
- Get the file **interest-groups** in the directory *netinfo* by anonymous FTP from **ftp.sri.com**.
- If you have World Wide Web access, try this URL address: **http://www.neosoft.com/internet/paml/**. Maintained by Stephanie da Silva, it lists and groups many mailing lists.

ELECTRONIC NEWSLETTERS AND JOURNALS

There is nothing "unique" about electronic newsletters and journals in the way that there is with USENET newsgroups and electronic mailing lists. It's just that increasing numbers of these newsletters and journals are no longer being published in print media and exist in electronic form only. Access to these electronic-only periodicals varies widely. Some you subscribe to; they come weekly, monthly, or quarterly in your e-mail (some are free; some you pay for); others you can get by file transfer (FTP) from Internet computers where they are archived; and still others you can read online through the World Wide Web.

How do you find out what electronic newsletters and journals related to engineering exist? Send e-mail to **listserv@acadvm1.uottawa.ca**, leave the subject line blank, and in the body of the message write **GET EJOURNL1 DIRECTORY** on one line and then on the next line, write **GET EJOURNL2 DIRECTORY**.

Note For general information on USENET newsgroups, mailing lists, electronic newsletters and journals, see Paul Gilster's *The Internet Navigator.* Check the specific chapters on these topics, but also check the directory of the Internet resources. Also take a look at the *New Rider's Official Internet Yellow Pages.* This book attempts to index electronic newsgroups, mailing lists, newsletters, journals, and the other Internet resources by the fields or subjects they cover. (See the end of this chapter for bibliographic details.)

INTERNET SEARCH TOOLS: ARCHIE, GOPHER, WAIS

To find resources on your own, the Internet provides a number of search tools.

Archie. You can use the Archie program to find things on the Internet; check any of the commercial books on the Internet. Archie enables you to search for files on the Internet containing a word or phrase. For example, you could search for "civil engineering" (which will create a deluge) or "solar-powered automobiles" (which might get you little or nothing). Archie gives you a listing of file names and their locations. Then you have to go out and FTP those files to see if they are what you want.

Gopher. Another resource to know about is Gopher. This is an Internet-based menuing system used by organizations, mostly universities, around the world. Engineering schools, libraries, societies, or projects have set up gopher menus to all their online resources. You'll find things like academic programs, faculty listings, project data, and occasionally collections of reports. If you knew that work was being done on your topic at a certain university, you could check to see if that university had a gopher service. You could do this by using gopher menus that are arranged by nations, states, cities, universities, and so on. You can also search all the gophers in the world for your topic using Veronica and Jughead. These search "gopherspace" for menu titles bearing your search terms. If you were to use Veronica to search for "civil engineering," you'd get a listing of all the gopher menu items in the world that had information on that topic.

WAIS. A further evolution in Internet-based search tools is WAIS (Wide Area Information Server). This search tool makes searching much easier and gives you easier access to the things you find. And it is not limited to FTP files or gopher menu titles. Anything made available to it is searchable, not just titles but words within the text of articles, names of mailing lists and newsgroups, and so on. You could search for electronic newsgroups, mailing lists, newsletters, and journals using WAIS.

WORLD WIDE WEB

The Internet world has been revolutionized by the World Wide Web, or the "web" for short. Almost all of the access and search techniques previously discussed are integrated and consolidated by the web. Not only can you search but you can view, and not just view text but also graphics. Using the web is much easier than using Archie, Veronica, Gopher, Telnet and WAIS. You just move your mouse pointer around and click on what you want. (If you don't have a web browser or access to the web, you can use a program like LYNX, which is a text-only approach. You still get to link, but you don't see the graphics.)

With the web, you are still limited to whatever people feel like making available on it. Quite a number of engineering resources exist on the web as the following discussion will show, but there is no overall plan. Some might list their engineering courses; others, their engineering faculty; some might post some locally written reports on a specific project. Some might have an online newsletter for a particular engineering society. But it's not like an organized library. Still, there are fascinating resources out on the net and particularly on the web. The difference is that it takes knowhow, energy, and patience to get to them.

World-Wide Web Virtual Library. The originators of the web have developed a menu of links to information sources in many different areas and fields. Use the web address **http://www.w3.org/hypertext/DataSources/bySubject/Overview.html**. When you get there, you'll see links to aeronautics and aeronautical engineering, biotechnology, standards and standardization bodies, and of course engineering. The engineering link takes you to another web page full of links to engineering fields such as civil engineering, mechanical engineering, and materials engineering. You'll also find links to engineering-related societies, journals, newsletters, design and research centers, academic institutions, special projects, plus good tips on finding other online engineering information.

Whole Internet Catalog. The O'Reilly publishers also provide a useful list of resources on the web. Use the web address **http://nearnet.gnn.com/wic/newrescat.toc.html** to get to the main page, and under the heading Science & Technology click on Engineering. There, you'll find links to the American Society for Engineering Education, the Cornell Theory Center Server, Electronic Engineering Times (an electronic journal), IndustryNET, International Organization for Standardization (ISO), National Institute of Standards & Technology, The Semiconductor Subway, and other resources as well.

Internet Connections for Engineering. The engineering library at Cornell University has developed a web page for engineering-related resources. Use the web address **http://www.englib.cornell.edu/ice/ice-index.html** to get to it. You'll see links to resources for most of the engineering fields plus links to employment resources, libraries, journals, patents, standards, technical reports, and so on.

Gopher Jewels—Engineering. "Gopher Jewels" are collections of related gopher menus arranged for easy access, and they can be accessed on the web. Use **http://galaxy.einet.net/GJ/index.html** to get to the home page for this collection, or use **http://galaxy.einet.net/GJ/engineering.html** to go directly to the resources for engineering. There, you'll see things like engineering collections of the Australian Defence Force Academy, North Carolina State University, Penn State University Libraries, RiceInfo (Rice University CWIS), as well as links to the Cornell Engineering Library Gopher, Electronic Journals for Engineering (CIC.net), Institute of Electrical and Electronics Engineers, NIST Electronics and Electrical Engineering Laboratory, and the University of Michigan, College of Engineering, Technology Transfer.

Note Most of the engineering resources in the World Wide Web refer to each other. Once you get moving on the web, you'll find links to most of the other engineering resources. Also, most of these resources have keyword-search capability. Type your engineering field or topic into these search dialogs, and see what else you can find.

World Wide Web Search Tools. How can you survey the entire World Wide Web for engineering resources related to your topic? Several tools exist for searching the web: World Wide Web Worm, Lycos, WebCrawler, JumpStation, and Web Nomad. Check any commercial book on the web for their addresses. There are some web sites that can help you find these search tools. Try the following addresses; if they have changed, look for "unified search engines" in books on the web:

Meta Index: **http://cui_www.unige.ch/meta-index.html** (produced by Oscar Nierstrasz)

SUSI: **http://web.nexor.co.uk/susi/susi.html** (produced by Martijn Koster)

Twente University's External Info: **http://www_is.cs.utwente.nl:8080/cgibin/local/nph-susi.pl** (produced by Twente University in the Netherlands).

EXERCISES

If you are not familiar with library-based information sources, find a technical topic that is of interest to you, and look for information related to it in as many of the following sources as you can:

1. Check the catalog at your library and, if possible, one of the Internet-accessible libraries mentioned in this chapter. For the three most useful-looking books related to your topic, make a bibliographic entry using the format shown in the documentation section of Chapter 6.

2. Using *Ulrich's International Periodicals Directory* or some other journal-locating resource mentioned in this chapter, find three useful-looking journals related to your topic. Make a bibliographic entry for each one.
3. Using one of the periodical indexes discussed in this chapter, find three useful-looking articles related to your topic in technical journals. Make a bibliographic entry for each one, again using the format shown in Chapter 6.
4. Using *NTIS* or some similar resource mentioned in the technical reports section of this chapter, find three technical reports related to your topic, and make a bibliographic entry for each.
5. Using one of the patent indexing resources discussed in this chapter, find at least one patent related to your topic, and make a copy of the record.
6. Using one of the catalogs described in the product literature section of this chapter, find at least one company involved with products or services related to your topic, and make a bibliographic entry for it.
7. If you have access to the Internet and to the World Wide Web, attempt to find one electronic mailing list, one USENET newsgroup, and one web site (on the World Wide Web) related to your topic. Use the search tools available on the Internet and the World Wide Web to assist you in these searches.

BIBLIOGRAPHY

Ardis, Susan B. *A Guide to the Literature of Electrical and Electronics Engineering.* Littleton, CO: Libraries Unlimited, Inc., 1987.

December, John, and Neil Randall. *The World Wide Web Unleashed.* Indianapolis, IN: Sams, 1994.

Fischer, Martin A. *Engineering Specifications Writing Guide.* Englewood Cliffs, NJ: Prentice-Hall, 1983.

Gilster, Paul. *The Internet Navigator.* New York, NY: Wiley, 1994.

Kamm, Lawrence L. *Successful Engineering: A Guide to Achieving Your Career Goals.* New York, NY: McGraw-Hill, 1989, p. 145.

Krol, Ed. *The Whole Internet: User's Guide and Catalog.* Sebastapol, CA: O'Reilly, 1994.

Maxwell, Christine, et al. *New Riders' Official Internet Yellow Pages.* Indianapolis, IN: New Riders Publishing, 1994.

Schenk, Margaret T., and James K. Webster. *What Every Engineer Should Know About Engineering Information Resources.* New York, NY: Marcel Dekker, Inc., 1984.

Subramanyam, Krishna. *Scientific and Technical Information Resources.* New York: Marcel Dekker, Inc.,1981.

Thomas, Brian J. *The Internet For Scientists and Engineers: Online Tools and Resources.* New York: IEEE Press, 1995.

Wiggins, Richard W. *The Internet For Everyone—a Guide For Users and Providers.* New York, NY: McGraw-Hill, 1994.

8

ENGINEERING YOUR PRESENTATIONS

Engineers are often asked to speak, and many engineers find they speak a lot. Whether you give an impromptu 5-minute briefing or a formal 1-hour presentation at a technical seminar (or something in between), you should see your talk as a great opportunity to share information and to show that you know how to communicate. Few of us are naturally gifted with such capabilities, and some of us are almost petrified at the thought of talking before a group, but the skills possessed by good speakers can be learned. The principles discussed in this chapter will enable you to become a confident speaker people will listen to, because you transfer information efficiently and effectively—that is, with a minimum of noise.

PREPARING THE PRESENTATION

Developing a worthwhile presentation is like developing a product: research and planning are crucial in the early stages. We all know what it's like to have to come up with a spontaneous briefing or unexpected oral report, but fortunately we usually have some lead-in time before we talk. Using that time to work through the procedures that follow will help you design a successful presentation.

ANALYZE YOUR AUDIENCE

Much of what was said at the beginning of Chapter 2 about focusing on your reader and purpose *before* writing can be transferred to preparing for an oral presentation. We've all been bored by talks that were above our heads, too simplistic, or unrelated to our interests. To make sure you don't do the same thing to others, ask yourself the questions in Figure 8-1 when beginning to prepare your talk, and make sure you have as clear an idea of the answers as possible.

DECIDE ON YOUR PRIMARY PURPOSE

Your purpose in talking is intimately related to the makeup of your listeners and the reason they are sitting in front of you. Are they there for instruction, information, insight, to be persuaded, or what? What action or change do you feel they need to undertake? Knowing exactly what kind of assignment you have will also determine your foremost purpose. Engineering presentations can take many forms, as Figure 8-2 indicates, each with a specific purpose and organizational requirements.

Make sure you know what you are getting into, what is expected of you, who your audience is going to be, and what you want to accomplish by talking to them. Decide exactly what you want your listeners to take away from your talk. Then you will be on solid ground while preparing the remaining features of your presentation.

DETERMINE YOUR TIME FRAME

The cardinal rule here is *never* to speak longer than you are supposed to. To avoid annoying busy people or offending speakers who come after you, check how much time you have been allotted. Knowing your time limit will also help you decide how much detail you can go into, how much time you should allow

- Who will the key individuals in my audience be?
- What needs or concerns do they have regarding my topic?
- How knowledgeable are they about my subject?
- How can I get their attention and interest right away—and keep it?
- What are their attitudes likely to be regarding what I have to say?
- Do I need to work on changing their attitudes, and if so, what is the best way to go about it?
- What benefits are they going to get from listening to me?
- What kinds of questions are they likely to ask?

Figure 8-1 Some questions to ask about your listeners *before* you give your talk.

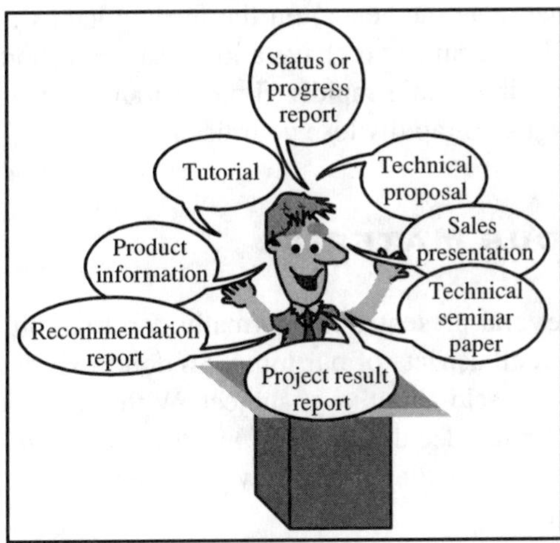

Figure 8-2 Just a few of the many kinds of presentations engineers find themselves giving.

for questions or discussion, and how much time you can spend on an introduction and conclusions or recommendations if have you some.

As Figure 8-3 illustrates, how deeply you go into different aspects of a typical engineering topic is related to how much time you have to speak. The tops of the pyramids in the figure represent the least you could say on a topic—perhaps a single sentence—while the true base of the pyramid (unseen in the illustration) represents everything that could possibly be said. This is perhaps why we almost always impose time limits on speakers!

It has been claimed that any subject can be covered in virtually any amount of time. A speaker could compress the creation of the universe into three or four sentences, or fewer, if necessary. In the same way, an expert could talk for hours

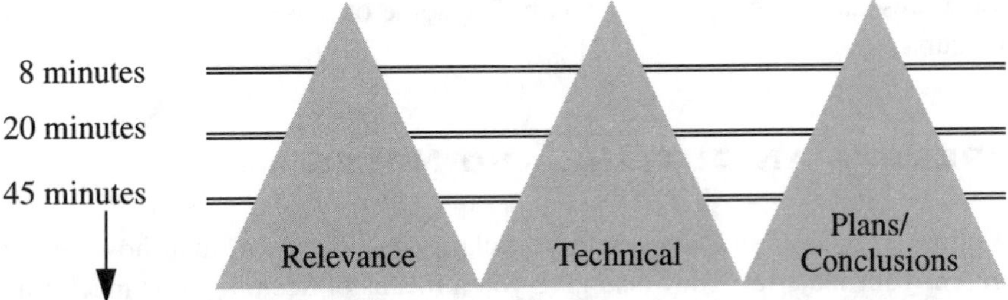

Figure 8-3 No matter how little time you have to talk, you can say something worthwhile in an engineering presentation. You just have to correlate how deeply you go with the time you have.

(probably to a rather small audience) on the feeding habits of fleas. Whatever you do, don't decide that since you have a lot to say in a short time you should speak as fast as possible while rapidly flipping transparencies on and off the projector. You will just as rapidly lose your listeners.

CONTROL YOUR MATERIAL

As indicated above, oral presentations normally don't permit us to give all the minute details about all aspects of our topic. So don't expect to say everything that could possibly be said about your subject. With a clear idea of your main purpose and time frame, decide *what the most important points are* that you want to get across to your audience and how you want to develop those points in the time you've got.

You may need to be quite mathematical about this. If you have 20 minutes to make four important points, you may subtract the time you want for an introduction and conclusions and divide what is left by four, thus leaving from 3 to 4 minutes for each point. In some presentations you may not want to give equal time to each point. For instance if you have to discuss five reasons why a project should be canceled or why your company should invest in new equipment, you might determine which points will meet the most resistance from your audience. Then aim to spend more time explaining those points while giving briefer treatment to the less controversial ones.

CHOOSE AN ORGANIZATIONAL PLAN

Your subject and purpose, and to some extent the time you have, will help determine how to organize your material. Most presentations can roughly be broken down into an introduction, the main points, a conclusion, and a question and answer period. Table 8-1 is a list of ways to organize your central material. Some presentations will call for combining some of these organizational plans, of course.

PREPARE AN OUTLINE AND NOTES

Writing an outline and notes help you clarify in your own mind how best to present your material. They also give you a means of deciding how much time to allot to each point and will be helpful when you rehearse the presentation. However, extensively relying on an outline or notes during the actual presentation can be dangerous as you will give the impression that you don't know your topic thoroughly.

TABLE 8-1 Examples of Ways You Can Organize Your Material for an Effective Oral Presentation

ORGANIZE YOUR MATERIAL IN THIS FORMAT...	IF, FOR EXAMPLE, YOU WANT TO...
1. Time sequence	Describe progress on a project or steps in a procedure Relate decisions leading up to an action or occurrences that led to a problem
2. Spatial sequence	Describe a piece of equipment or a physical area such as a test site or plant facilities Outline a physical process
3. Decreasing importance	Give your most important points first down to the least important: Relating six ways to improve or prevent a situation
4. Increasing importance	Work up to your most important point: some minor reasons for an action, change, or decision, followed by the major reasons
5. General to specific	Present a general point followed by specific examples: "We've got to improve production," followed by concrete ways to do so
6. Specific to general	Be persuasive: Cite examples of personal injury to lead to the point that more stringent safety regulations are needed at your plant
7. Comparative	Compare and contrast equipment, approaches, or ideas on such aspects as costs, durability, reliability, ease of operation, etc.
8. Familiarity	Begin with the familiar first, leading your audience into an understanding of the unfamiliar: Talking about corporate needs or problems
9. Difficulty	Present data in order of the easiest first and progressing to the hardest, as in a training session or tutorial
10. Controversiality	Begin with least controversial points in order to be diplomatic about sensitive issues: Why changes should be made in a project in which people have some ego investment

Your notes and outline may range from a few hastily scribbled ideas—if that much—jotted down a few moments before an unexpected briefing, to a complete manuscript of every word you intend to say. Reading a word-for-word written version of your talk in front of an actual audience is NOT a good idea, however, unless you feel very insecure or are giving a highly technical conference paper calling for extreme precision and accuracy. Even then, few people want to sit while a paper is read to them; after all, they could read it themselves in the comfort of their own homes or offices.

While preparing your talk, determine which prompts will best keep you on track when giving it. The main cues engineers tend to use are the following:

- An **outline** of the complete talk, with key ideas highlighted or in large print to be quickly glanced at if necessary as the presentation goes along (see Figure 8-4).
- **Note cards** numbered in the order they will be used, with key ideas and facts clearly written on them. If these are relied on too much, however, you will appear unsure of your material.
- **Visual aids,** such as transparencies or slides. If you are really on top of your topic, the visuals themselves will be all you need to keep on track. They may in fact be the outline of your talk. If you wish, you can make printed copies

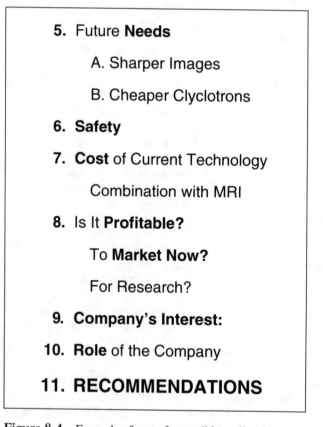

5. Future **Needs**

    A. Sharper Images

    B. Cheaper Clyclotrons

6. **Safety**

7. **Cost** of Current Technology

    Combination with MRI

8. Is It **Profitable?**

    To **Market Now?**

    For Research?

9. **Company's Interest:**

10. **Role** of the Company

11. **RECOMMENDATIONS**

Figure 8-4 Example of part of a possible outline for an oral presentation. Note that key points are highlighted. In an actual outline you might want to spread the items farther apart to be easily noted by a quick glance while you talk.

of them for your own use, with notes written to yourself that can be quickly referred to if needed.

CREATE SUPPORTING GRAPHICS

Because we live in an increasingly visual age, and since people remember information better when they both hear it and see it, most effective engineering speakers support their talks with illustrations of some kind. Graphics also work to your advantage since preparing them forces you to organize and rehearse your presentation and possibly discover weak spots that need attention. Showing them will save you time during the presentation because you won't have to write the information down on a board or flip chart. They can also serve as cues for you, reminding you of what you want to cover and the order in which you want to cover it. You should at least plan to use visuals wherever you feel they will

- simplify a point
- clarify a point
- stress a point
- show critical relationships between ideas or facts

Channels for Graphic Support. For engineers, the most common means of showing graphics is the overhead projector with *transparencies* (also known as foils, overheads, visuals, or view graphs) which you prepare in advance. Transparencies also have the advantage of letting you add information on them with a wax marker while showing them, or even using additional ones as overlays. If you have several transparencies, remember they can be slippery. It's embarrassing to see them slide off the table and across the floor, so you might want to consider a matting of some sort for each one.

Slides shown from a *slide projector* can be effective also, especially if you can do a professional job with color and art work, but don't overdo it. It's also possible to display visuals directly from a personal or laptop *computer* connected to a projector, showing the audience what you have created on a graphics program and allowing you to progress through your talk by calling up graphics through your keyboard. A *flip chart* is useful for ongoing illustrations or emphasis during your talk, as is a *chalkboard*.

If you are talking about a specific piece of equipment that you can bring into the room with you, do so—as long as your audience is small and close enough to be able to see it.

Designing Your Graphics. Any good graphics program will give you everything you need to create graphs, pie charts, bar charts, flow charts, etc. Nowa-

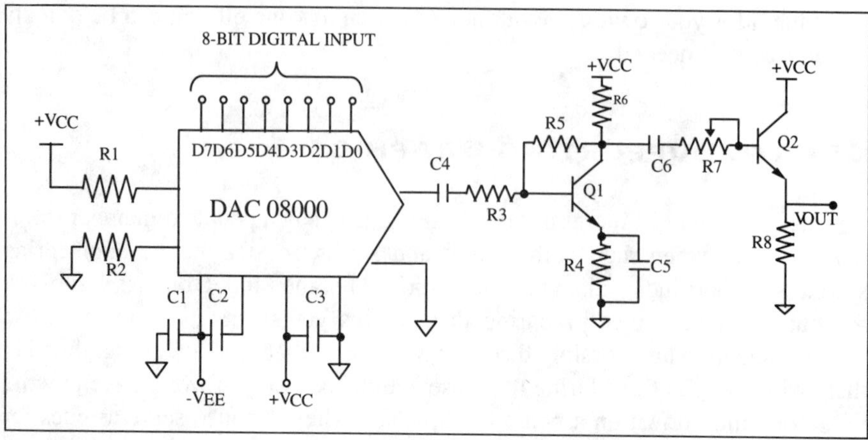

Figure 8-5*a* An example of an overcrowded transparency. Far too much information is thrust upon the audience here. One way to make this material more accessible would be to reduce the circuit first to a block diagram, as shown in Figure 8-5*b*, and then, if more detail is needed, expand the drawing one block at a time on separate visuals, as in Figure 8-5*c*.

days with only a few hours' training any engineer can produce almost any kind of graphic with programs like Harvard Graphics, MacDraw, MacPaint, Lotus Graphics, Power Point, and many others. Some specialized programs allow you to create excellent illustrations of equations, electrical circuits, and other technical data. With the growing use of scanners you can now copy and present professional illustrations or photographs of just about anything. If you use a scanner, be aware of any copyright restrictions and give credit to the source of any scanned material.

Remember, the most dazzling transparency or slide will impress no one if the information it contains is not easily accessible. In fact anything you put on the screen that cannot readily be grasped by your audience—because it's either (a) **too complex** or (b) **too small**—is almost worthless. This point is particularly worth heeding if only because we see it so often ignored by engineering speakers.

a. Don't let your visuals suffer from **information overload.** Each should be as simple as possible, portraying the bare facts—you can elaborate verbally. Even quite technical material can be reduced to manageable concepts on a screen. For example an electronics circuit can be broken up into constituent parts after a simplified overview has been given, as shown in Figures 8-5 *a–c*.

b. If your visuals consist of lists or other written information, **make the words easy to read** (see Figure 8-6). This means using at least a 24-point

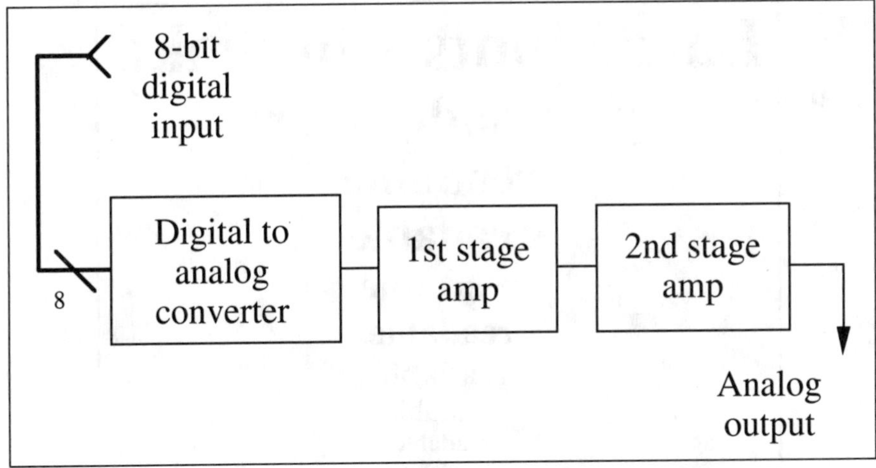

Figure 8-5*b* Simplified block diagram of the circuit in Figure 8-5*a*.

font size, preferably boldface. It's best to have no more than eight lines of print on a slide or transparency, and better not to use all capital letters since this makes for harder reading. A page of text or an illustration photocopied from a book or journal rarely makes a good overhead.

When you present written information on the screen, don't crowd it. Provide ample margins and plenty of white space between and around the lines.

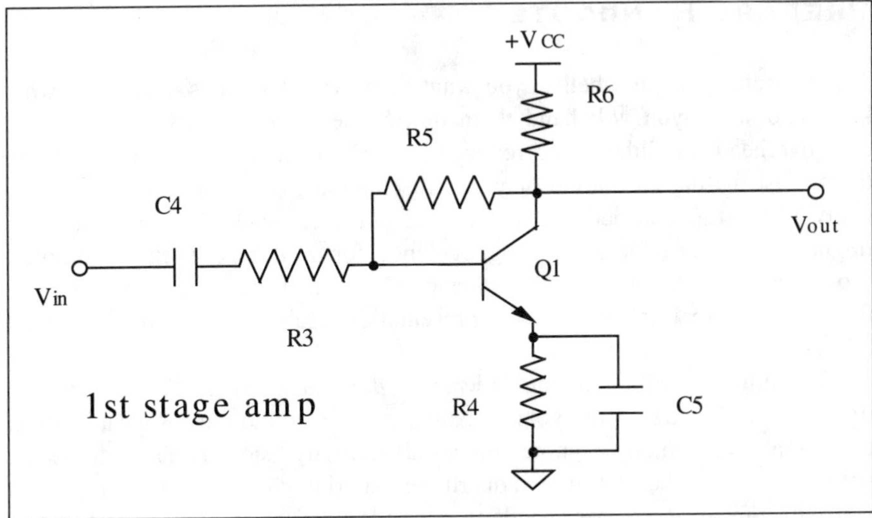

Figure 8-5*c* The center block from the diagram expanded to part of the original circuit.

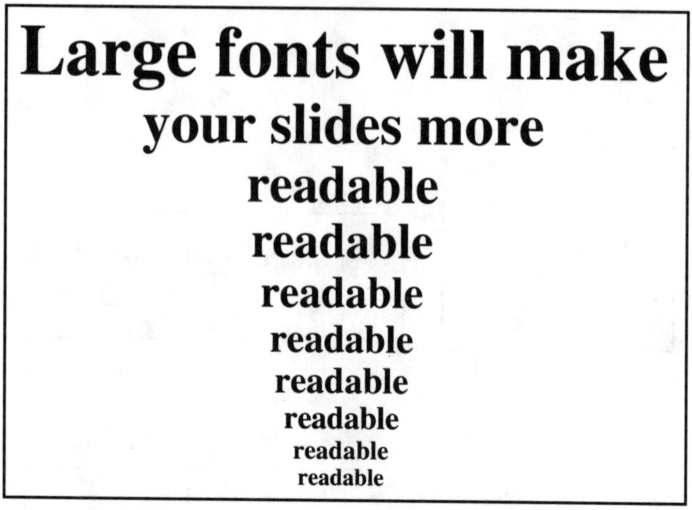

Large fonts will make
your slides more
readable
readable
readable
readable
readable
readable
readable
readable

Figure 8-6 Use at least 24-point print on your overheads if you want your audience to read them.

You might want to use bullets, checks, or other marks to emphasize points, but resist the temptation to go overboard with the variety of fonts and clip art now available. Don't let your art work overwhelm the information on the screen.

PREPARE HANDOUTS

Think carefully about whether you want to provide handouts, and if so, what kind, and when you will hand them out. Some speakers pass out copies of their overheads or slides, often reduced in size, so listeners can make notes on them. Distributing an outline of your presentation may be a good idea, especially if the topic is detailed and covers a lot of material. Sometimes you might need to provide supporting evidence for your talk, such as samples, brochures or other data. On the other hand, a lot of successful speakers ask their listeners to just focus on the presentation itself, and take notes as they wish.

The dilemma with handouts is *when to pass them out.* If it's at the wrong time, they will distract from your talk since people tend to look at what is given them right away, and may ignore you or only partially listen. You need to decide beforehand when the best time to distribute material will be, and how and when you will refer to it, so that it adds to your listeners' concentration and understanding rather than detracts from it.

PREPARE YOUR INTRODUCTION

When thinking of how to begin your talk remember that

1. your audience may be asking themselves, at least subconsciously, "Why do I need to hear this?" or "Why should I be here right now?"
2. your audience has a limited attention span.

To help solve both problems, design your introduction to let your audience know right away what your topic is and of what benefit it is to them. Then provide a sense of direction by giving an overview of where you're going in the presentation and what you plan to cover. Let your audience know how long you intend to speak if it's not already known. Many speakers lose their audience right away because they fail to follow these procedures at the outset of their presentation.

PREPARE YOUR CONCLUSION

Design the end of your presentation to focus the audience's attention solely on essentials. Depending on the type of presentation you give you will be able to reinforce your message by, for example,

- summarizing what you have discussed
- stressing your central idea once more
- reviewing your key points
- restating your recommendations or decisions

An appropriate final graphic can be a great help here. Above all, give your audience a lasting impression of what you want them to take away from your talk, such as the feeling that you have solved a problem or concern, or have provided new insights. Don't suddenly stop talking at this point, however. Make a note to close gracefully with something like "And this concludes my presentation. Thank you for your attention. Are there any questions?"

GET READY FOR QUESTIONS

If there is going to be a question period after your talk, spend some time during the preparation stage to anticipate and get ready for them. Put yourself in the place of your listeners: Are they likely to find any part of your talk especially difficult, detailed, or controversial? Are they likely to hold any opposing viewpoints? Are there areas you may not be able to go into as thoroughly as you would like, due to time restraints, and which might therefore generate questions? What could be the "worst" question asked? Also, can you think of diplomatic ways to encourage questions from people who are reluctant to ask? (Sometimes a friendly smile or "I'd *love* to have some questions" is all that's needed.)

Remind yourself at this point that when questions come, it helps to repeat them in some form before answering. The repetition is useful for people who didn't hear the question clearly in the first place and the delay might give you time to gather your thoughts.

PRACTICE, PRACTICE, PRACTICE

Keep in mind the initials PPPPPP: Plentiful Practice Prevents Painfully Poor Presentations. Some speakers go over their material—outline, notes, visuals—up to seven times before presenting it; for others this would be overkill. Most speakers rehearse at least twice, however, if the talk is of any significance or if they feel unsure of their material.

Depending on the importance of your talk, you may decide to have at least one dress rehearsal, preferably in the room where you'll be presenting. This will also let you get familiar with the room and any equipment to be used. An audio or videotape of this rehearsal, if possible, would enable you to self-critique your performance. On the other hand you might want to find a trial audience to listen to your first run and give you some feedback, whether friends, colleagues, a spouse, or even yourself in the bathroom mirror.

Perhaps the most valuable outcome of careful practice is the self-confidence you gain. One antidote to nervousness about speaking in front of a group is to be able to walk into that room knowing you're completely in control of your subject and ready to present it in an effective manner—confidence you can gain by first practicing as much as possible.

DELIVERING THE PRESENTATION

All your preparation efforts are aimed at one goal: to give an effective presentation that will produce the desired results. By the time you stand in front of your audience you should have fixed many of the potential glitches that can surface in oral presentations. Knowing your subject and audience make-up will have helped you determine the information you need and how you need to communicate it.

Most engineers can prepare a presentation well enough given a little awareness, analysis, and preparation time, yet the sad fact remains that plenty of lackluster and somewhat boring presentations occur every day in business and industry. As with a written report, such presentations could be greatly improved by the elimination of noise.

AVOIDING NOISE IN ENGINEERING PRESENTATIONS

In an oral presentation, noise can be defined as anything that prevents the message from effectively getting into the minds of the audience. Following are some causes of noise that frequently occur in engineering (and other) presentations.

1. **Speaking too softly.** A common problem with beginning speakers is a tendency to speak too softly. Ironically, such softness is a form of noise since it prevents the message from clearly getting through to the listeners. You don't want to blast out your audience, but on the other hand you do want everyone to hear you. Try to project your voice relative to the room and audience size. If some listeners can't hear you, you're wasting their time.
2. **Speaking too slowly or rapidly.** A slow, labored pace with too many pauses causes boredom and decreases your credibility. The reverse of this is to talk too rapidly, either because of nervousness or because you're so much more familiar with your topic than your audience is. Aim for a normal conversational speed, but remember that pausing and deliberately slowing down once in a while can help you stress important points.
3. **Speaking monotonously.** You may have heard stories of the dull college professor who dreamed he was giving a lecture only to wake up and find he was. How you talk often makes a bigger impression than flashy visuals and what you say combined. You could be explaining the never before revealed secrets of time travel and yet find few paying attention if you sound bored to death. Hypnotic monotony can be avoided by varying your pace and your pitch—by speaking the way most people do in lively and enthusiastic conversation.
4. **Using verbal fillers.** When a speaker needs to pause or is uncertain of what to say next, irritating and empty catchwords or phrases like *uh*, *umm*, *basically,* and *yunno* sometimes take over. Don't distract your audience by your UPM (umms per minute) rate. There's nothing wrong with being silent for a few moments while gathering your thoughts.
5. **Blocking the screen.** Too many engineers stand directly in front of the screen and stay there throughout their talk, thus frustrating their neck-cran-

Figure 8-7 You put a lot of planning and work into your visuals, so don't create noise by standing between them and your audience.

ing audience and wasting any effort put into their graphics (see Figure 8-7). It's just as bad to stand partially off to the side yet still block the screen for those sitting at the sides. If possible, before your audience arrives, have a colleague sit in various seats and let you know where you should avoid standing for long. If you don't have time for this luxury, move around enough during your presentation to avoid blocking anyone's view continuously. Better yet, stand far enough to the side to prevent screen blockage from ever becoming a problem.

6. **Reading from the screen.** Generally avoid reading your slides during a presentation—they are aids but are not meant to be your entire speech. Straight reading takes time away from a deeper explanation of your topic, and may bore your listeners since they can read for themselves. Also, be wary of dimming the lights to make your visuals easier to read. Low lights can make people drowsy and can hide facial expressions and the eye contact you need to have with your listeners.

STRENGTHENING YOUR PRESENTATIONS

Use an Informal Style. When making an engineering presentation you're not delivering a sermon (usually) or pronouncing on a profound legal intricacy. Generally the best style is an informal one, paralleling as closely as possible the normal conversational mode of everyday life. It's quite all right to use contractions (*it's, don't, couldn't,* etc.), even if you avoid them in formal writing. Using pronouns such as *you, your, I,* and *we*, will help you relate to your audience's interests and needs, and will indicate you are interested in them as people rather than as an impersonal mass. Avoid long, complex sentences, substandard grammar ("Me and Jim here aren't no experts, but . . ."), and any technical jargon not readily understood by your listeners.

Make Clear Transitions. You may have a well-organized speech full of important details, but you could still lose or confuse your audience if you don't show the connections between your ideas. The key is to be explicit. Tell your audience when you're moving on to another aspect of your subject or are about to give an example. Your visuals will assist you, of course, but make your dialog as user friendly as possible. Keep your listeners in the picture by emphasizing connections and transitions in your thinking through the use of simple words and phrases like these:

- First
- Next
- To begin with
- Initially
- Furthermore
- Consequently
- As a result
- On the other hand
- As you can see
- For example
- Also
- Finally
- In conclusion
- To sum up

When you overlook such transitions in a written report your readers can at least go back over the material a few times and try to figure out the connections for themselves, exasperating as that might be. Someone listening to a talk has no such opportunity.

Repeat Key Points. No matter how brief the presentation, you're going to have at least one main point you want your audience to go away with. Don't be afraid to repeat yourself—your listeners need to know what you consider the most important aspects of your subject.

Given the sad fact that even the best of speakers may have someone in the audience whose attention strays, there is a lot to be said for that old piece of advice to "tell your listeners what you're going to tell them, then tell them, and finally tell them what you have told them." As we pointed out earlier in the section on preparing conclusions, it's essential that you repeat your key points in a concluding summary.

Use a Pointer. A pointer is the best way to focus your audience's attention on your graphics while you explain what they're looking at. Be careful not to overuse it, however, or to unconsciously wave it around when not pointing. Some people have a distracting tendency to open and close a pointer repeatedly due to nervousness, or to even scratch themselves with it. When not actually using it, keep your pointer firmly clasped in one or both hands.

The various kinds of pointers come with both advantages and disadvantages. A *straight metal* or *wooden stick* is easily available, but you have to stand fairly close to the screen to use it. This limitation can overly restrict your movement and may also cause you to block the view for some people. The *retractable stick pointer* is also easily available but has the same potential drawbacks. If you use a pointer, hold it with the arm closest to the screen so you don't have to turn away from the audience every time you point (see Figures 8-8 *a*, *b*).

The *laser pen* is a fairly new type of pointer which projects a red dot onto the screen and can be aimed from anywhere in a room. Its only drawback is that you must be extremely careful where you are using it because you don't want to permanently injure anyone's eye by directing the beam at them. Hold it firmly at your side when pointing if you are at all nervous—an unsteady hand will cause the laser's dot to dart around surprisingly.

Maintain Eye Contact. You increase your credibility a great deal through eye contact with your audience as you talk. While avoiding eye contact could give the impression that you're shifty or unprepared, looking at your audience helps establish rapport with each member in a small group and a sense of intimacy with a larger group.

Start off by making eye contact with the friendliest faces and most attentive people in the audience. As you progress in your talk try to hold visual contact with each person for a few seconds and move on to someone else. Looking at individuals also enables you to pick up feedback on how they are receiving your message; puzzled looks or frantic note-taking, for example, might show that you need to go back over something or slow down.

Be Ready for Unexpected Questions. You can prepare all day for questions and still land at least one question you never dreamed of. Try not to appear startled or surprised when this happens—you've prepared a lot and know your subject well. Two strategies for tackling unexpected questions are as follows:

1. Simply say you don't know. People will respect you for being honest, and you can still offer to supply possible sources for the information.

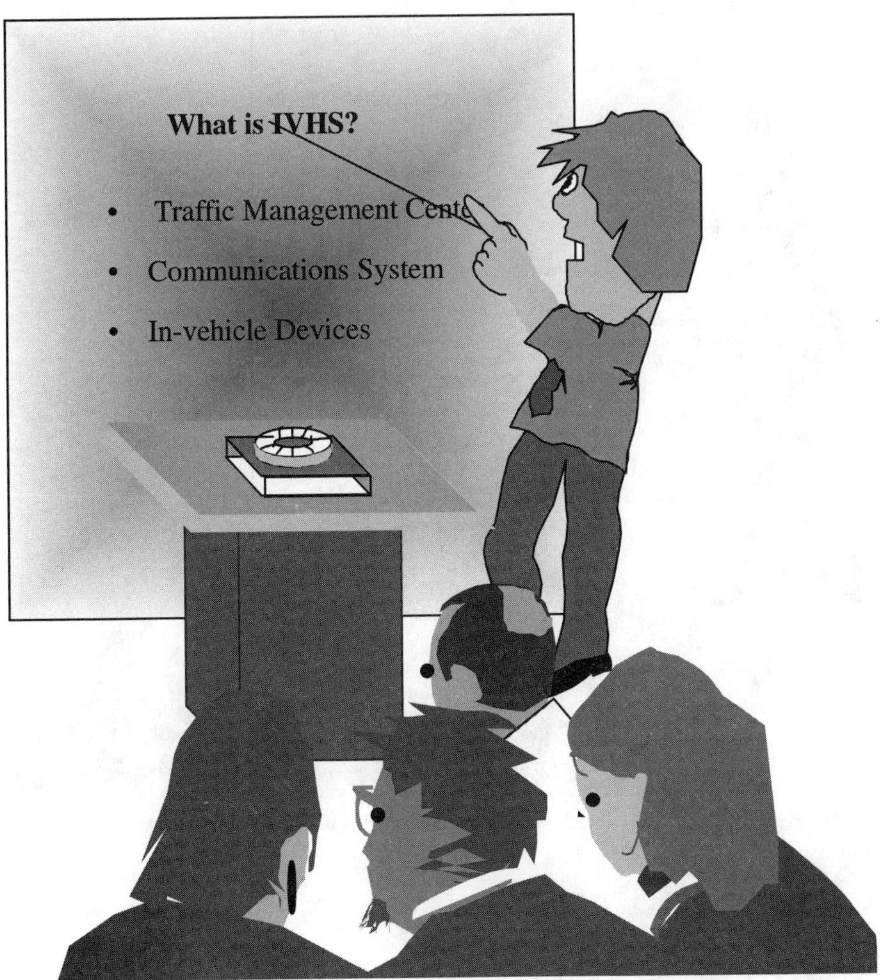

Figure 8-8*a* Using the arm farthest from the screen to point pulls you from eye contact with your audience.

2. Offer to talk with the questioner after your presentation. This may be the best answer if (a) the question is too involved for the discussion you are in, (b) the question is not really related to your topic, or (c) the question is hostile and you don't want to get into an argument. Rarely will anyone seek you out afterwards unless they are genuinely interested in information you may have.

Accept Your Nervousness. Although this fact may not be very encouraging if you're just starting to give presentations, the best cure for nervousness is experience. Until you have this experience, accept your nervousness as perfectly

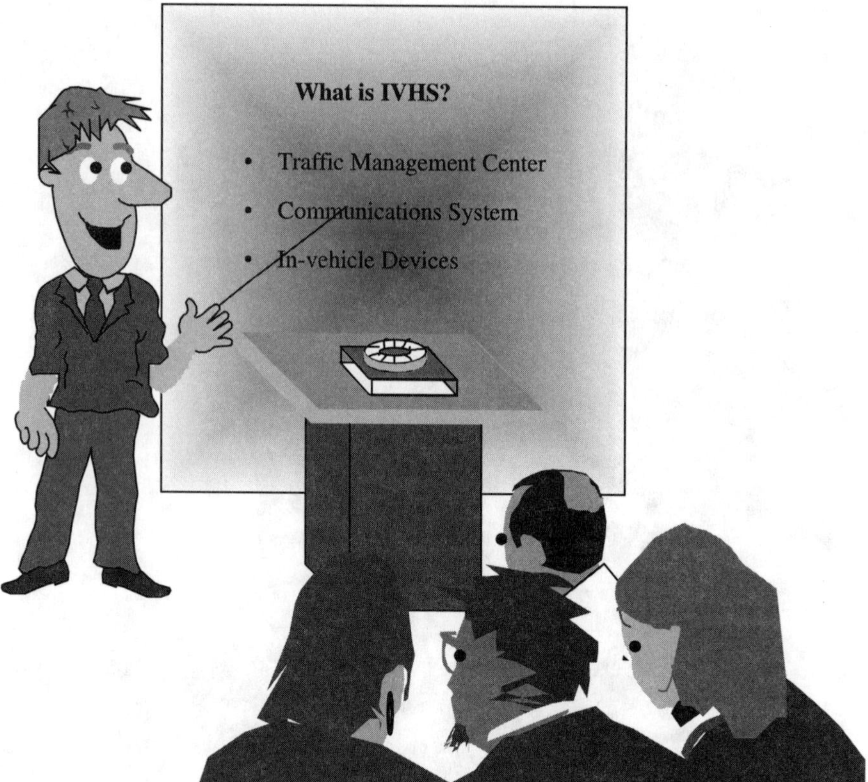

Figure 8-8*b* Using the arm closest to the screen still allows you to talk while facing your listeners.

normal. We all suffer from it (although our nervousness is often much less noticeable to others than we might think). Even the most experienced speaker sometimes gets tense before facing an audience; some learn to use their anxiety as a positive energizing power that helps them to be more alert and lively. If you have problems with stage fright, consider the following:

1. Enter the presentation room knowing you've worked hard on your talk and have practiced delivering it. In others words, give yourself as much reason as possible *beforehand* to be confident of your knowledge and ability. Then try to concentrate on your topic rather than on yourself. Appropriate gestures and facial expressions will often occur naturally in a well-prepared speech.
2. Take some deep breaths before entering the room. Even a short walk around the building or a few simple physical exercises may help relieve anxiety.

You can't continue with these activities once you're in the room, of course, and they can never substitute for the self-assurance that comes with really knowing your material.

3. Try to have a few friends or colleagues in the audience who can give you moral support through a reassuring smile or nod of the head. If everyone in the room is a stranger, look for a friendly face or two and exchange a few words of banter before launching into your presentation. This can relieve quite a bit of the tension you may be feeling.

TEAM PRESENTATIONS

Since engineers frequently collaborate on a project, compile a proposal, or report on a new product, you are likely to be involved in team presentations. These allow individuals to speak in turn on a topic, each with his or her own "part." Group presentations also permit a specific aspect of a complex subject to be presented by the individual who worked on it rather than by a team spokesperson. This kind of presentation is additionally effective because

- teamwork reduces everyone's preparation workload
- longer presentations are possible without exhausting one person
- speakers can enjoy team support during the presentation
- the variety of speakers helps hold the audience's attention
- each topic can be explained (and questions answered) expertly

PREPARING FOR A TEAM PRESENTATION

Whether giving the presentation with one partner or several, the first step is to get together to analyze your audience and purpose. Then decide on the main points to be stressed, the order in which the material will be covered, and who will cover which topic. The team leader should clearly partition the topics and make sure that each speaker sticks to the assigned topic and doesn't cover any other speaker's material.

It's essential to allocate time to each speaker early so that everyone can prepare accordingly and the presentation can conform to any required time limits. Decide beforehand whether questions from the audience should wait until after the entire presentation or should follow each speaker (if you have these options).

Because collaborating engineers do a lot of communicating with each other by rather impersonal memos or telephone messages, meeting to prepare the presentation may also provide an opportunity for all presenters to get to know one another better. If group members are familiar with each other, the presentation is likely to be more relaxed and look more polished and professional.

SHARING THE PRESENTATION

Assigning different parts of the presentation to alternate speakers prevents a long presentation from becoming monotonous, arouses more audience interest, and provides clear structure to the presentation. Plus, it's nice for each speaker to have a breather.

To ensure that your group presentation flows well, pay particular attention to how you are going to move from one speaker to the next. A simple lead-in like ". . . and now Eva is going to cover the financial aspects of the project" might be all you need. If there are just two of you, break down the topic so you can alternate back and forth for approximately the same total amount of time.

MAKING A DRY RUN

Groups, like individual speakers, need to put aside plenty of time for practice. Rehearsals will uncover any information gaps or neglected subtopics and ensure that all parts of your presentation are carefully merged. If team members have never performed together before, practice will be essential to ensure coordination in the presentation. As an individual team member you will also need to be certain you can conform to time limits: There is no better way to make enemies than to dominate the presentation and speak longer than you should, shrinking fellow speakers' time.

CHECKLIST FOR AN ORAL PRESENTATION

The items in Figure 8-9 can be used to evaluate an oral presentation if grades or scores are assigned to each item. You might prefer to use the list *before* you give your talk, however. Put yourself in the place of your audience and try to get a sense of how they would "grade" you as they listen to your presentation. You might even get a friend to check you out on each item during a dry run of your presentation.

INTRODUCTION

- ☐ Creates favorable atmosphere
- ☐ Creates appropriate pace
- ☐ Hooks listener's attention
- ☐ Relates subject to listeners
- ☐ Presents clear central idea

BODY

- ☐ Reveals careful audience analysis
- ☐ Supports central idea
- ☐ Maintains audience interest
- ☐ Provides technical accuracy
- ☐ Organizes details effectively
- ☐ Allocates time carefully
- ☐ Provides clear transitions

CONCLUSION

- ☐ Ties presentation together
- ☐ Restates central idea
- ☐ Proposes action or response
- ☐ Invites discussion or questions

ATTENTION TO TIME LIMITS

- ☐ Too short
- ☐ Just right
- ☐ Too long

VISUAL AIDS

- ☐ Are clear and easy to read
- ☐ Look professional
- ☐ Avoid information overload
- ☐ Clearly support related ideas
- ☐ Are enough

DELIVERY

Sound

- ☐ Clear volume and pronunciation
- ☐ Effective diction
- ☐ Varied speech patterns
- ☐ Absence of *uh-huh, y'know, basically . . .*
- ☐ Adequate enthusiasm
- ☐ Standard grammar and usage
- ☐ Good question response

Appearance

- ☐ Professional posture and appearance
- ☐ Appropriate gestures and mannerisms
- ☐ Effective use of pointer
- ☐ Consistent eye contact with audience
- ☐ Competent handling of notes and visuals

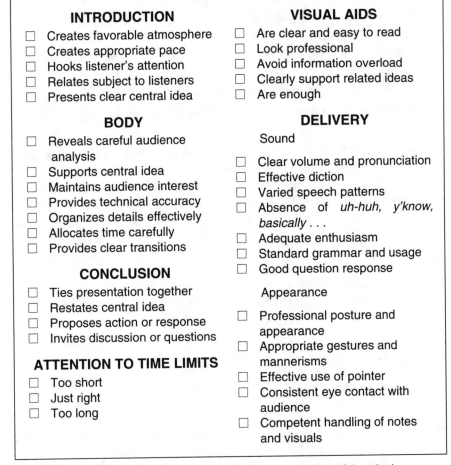

Figure 8-9 The aspects of an effective oral presentation. Even if they don't seem important to you, they will be to your audience.

LISTENING TO A PRESENTATION

You will probably listen to more presentations than you give during your engineering career, yet listening is the most neglected of all communication skills. As a speaker you may have already found there's nothing more frustrating than working long and hard on a speech just to have an unresponsive audience, and the point can be made that the responsibility for a successful presentation lies partly with the audience. Here are some ways to be a good listener:

1. Maintain natural eye contact with the speaker.
2. Show by your posture that you are alert, interested, and well-disposed toward the speaker.
3. As much as possible, ignore distractions such as people talking or other external noise.
4. Take notes on the speaker's most important points.
5. Develop at least one question in your mind, and ask it at the appropriate time.

In fact one of the best ways to be a good listener is to ask questions. A question lets the speaker know you're paying attention and that the presentation has made you think. Also, a question helps a presenter decide which information is most important and where the main concerns of the audience lie. Even if you understand everything presented, why not ask about something you found particularly interesting or new?

It's not hard to see how actively focusing on a speaker and concentrating on what is being said establishes a sense of empathy that leads to more efficient information transfer. Being an active listener, then, is not just a matter of being 'nice' to the speaker; it also rewards you with a more complete appreciation, knowledge, and evaluation of the material presented.

EXERCISES

1. Contact three or four engineers and find out what kinds of oral presentations they make on a fairly regular basis. How long do they talk each time? What do they talk about? Who is their audience? How do they meet the needs of the audience? Do they use any graphic aids or handouts while presenting? What kinds of feedback do they get? How do they know whether their presentation has been successful?
2. Listen to someone giving an oral presentation and evaluate his or her performance as best you can using the checklist in Figure 8-9. (You may need to be very diplomatic about this.) After evaluating the presentation, think of ways you might improve it if you had to give it yourself.
3. Take a written report—your own, if possible—and turn it into an oral presentation. What do you have to consider when preparing the talk that was not important when writing the document? Do you have to change the order or emphasis of any material? Do you find material in the written report that can be presented graphically in the talk? Does an outline of the report give you ideas of what to present graphically? How will you introduce and conclude your presentation? Does the fact that you may be asked questions after the talk cause you to think differently about your material than you did when writing the report?

4. Think of the various people you have listened to, such as teachers, fellow professionals, preachers, or politicians. What are some of the best things you have heard or seen these people do? What are some of the worst? Who were the most effective speakers? Who were the least effective? What can you learn from both kinds that will help you improve your own skills in giving oral presentations?

BIBLIOGRAPHY

Adamy, David. *Preparing and Delivering Effective Technical Presentations.* Norwood, MA: Artech House, Inc., 1987.

Beer, David F., ed. *Writing and Speaking in the Technology Professions.* New York: IEEE Press, 1992.

Emery, Blake, and Karen Klamm. "Effective Listening," *Proceedings of the 33rd International Technical Communication Conference*, pp. 129–131, Detroit, 1986.

Lucas, Stephen E. *The Art of Public Speaking*, 5th ed. New York: McGraw-Hill, Inc., 1995.

Mablekos, Cariole M. *Presentations That Work.* New York: IEEE Press, 1991.

Woelfle, Robert M., ed. *A New Guide for Better Technical Presentations.* New York: IEEE Press, 1992.

9

WRITING TO GET AN ENGINEERING JOB

Two tools commonly used to seek employment are the application letter and the resume.[1] You send one or both of these to prospective employers when you are in a job search. What combination depends on the potential employer and the advertisement—some request only the resume, some request only the letter, and some don't indicate. When you're not sure, send both.

HOW TO WRITE AN ENGINEERING RESUME

As you know, a resume is a summary of your professional experience, education, and other background relevant to the employment opportunity you are seeking. You can think of it as highlights on who you are professionally—a summary of your career to date.

The key to writing an effective resume, that is, one that highlights your best qualifications, is a design that can be scanned in about 20 to 30 seconds. The prospective employer ought to be able to glance at your resume for that period of time and have a good understanding of your qualifications and background.

CONTINUOUS REDESIGN AND UPDATE

When you develop a resume, it is not a one-time effort that you can subsequently forget. You may have to redesign it substantially for every new employ-

[1] Our special thanks to Randy Schrecengost, P. E., for his reviews and recommendations on this chapter.

ment opportunity you go after, and you must remember to update it at every accomplishment, milestone, or new phase in your career.

It used to be that people considered the resume a fairly permanent record of their background, updated only infrequently. They could use roughly the same resume for many different job searches over a number of years. However, with increasing competition in the job markets and with the availability of desktop publishing software, all that has changed. Now, many job seekers redesign their resumes for practically every new employment opportunity.

Redesign or no, you must be conscientious about updating your resume whenever you complete new projects, change assignments, add new skills to your professional repertoire, and, of course, whenever you change jobs. This constant updating is particularly important if you settle into one company for a long time. It's easy to forget details about what you've done professionally over the space of just 5 years. For that reason, keep a working draft of your resume always at hand, whether as a computer file or as a hardcopy printout on which you scribble notes whenever your career takes a new turn.

Note You should be aware of a trend toward electronic processing of resumes. Many companies now electronically scan the resumes they receive, searching for keywords relating to specific experience, qualifications, and skills that they need. This automated process generates interview invitations to those who have the right keywords and not-interested letters to those who do not. Thus, having specific detail in resumes in this context is critical. To permit scanning, electronic resumes must use simple type without bold, italics, underscores, or different fonts and typesizes. For more information on electronic resumes, see Joyce Kennedy and Thomas Morrow, *Electronic Resume Revolution.*

DESIGN COMPONENTS

The design of a resume is certainly important to success in an employment search. But a resume can't do it alone—many other elements have to be present such as connections, timing, need, and of course your actual qualifications. A well-designed resume does a number of things for you: It highlights your best qualifications and makes it easy for readers to see them quickly; and it conveys a sense of polished professionalism that reflects upon you.

Chronological or Functional Organization. One of the first issues in resume design is whether to divide your background information chronologically or thematically. To get a sense of these two organizational approaches to resumes, look at the illustrations in Figure 9-1 for a schematic view and Figure 9-2 for a full-content view.

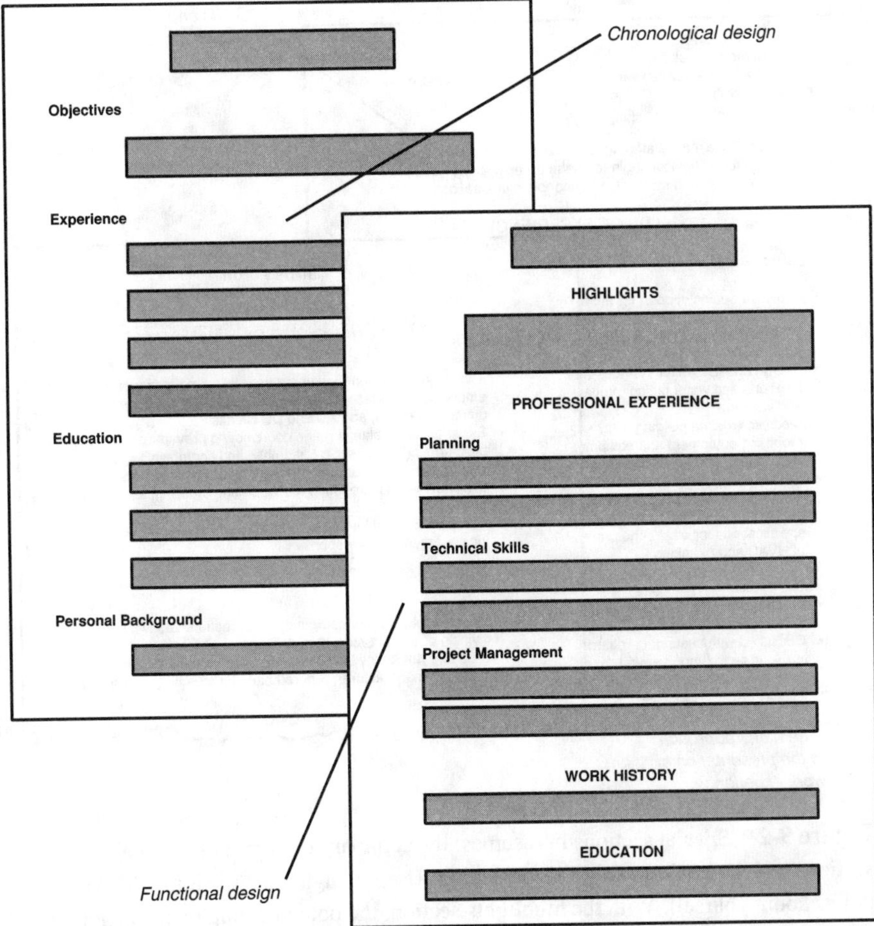

Figure 9-1 Schematic view of resume designs. Decide whether the chronological or the functional design works best for you. Visualize the headings you'll use and their relation to each other and to the body text.

- *Chronological approach*—divides your background into education, work experience, and possibly military (although military experience can be distributed into the education and experience sections).

  One of the strengths of the chronological design is that it shows your work history well. Also, it shows your responsibilities and projects for each organization you've been with. In the education section, this design shows where you studied and what you studied while there. However, the chronological design does not give a capsule picture of your key qualifications and your key strengths—that information is spread across the work and education sections. (A way to solve this problem is to add a highlights section, discussed later in this chapter.)

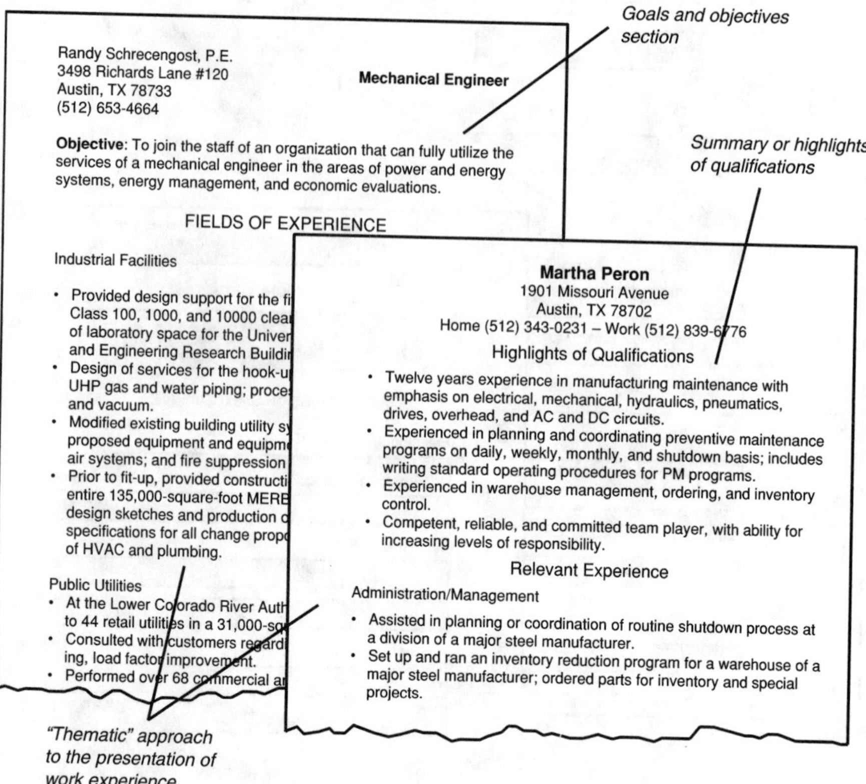

Goals and objectives section
Summary or highlights of qualifications
"Thematic" approach to the presentation of work experience.
Randy Schrecengost, P.E.
3498 Richards Lane #120
Austin, TX 78733
(512) 653-4664
Mechanical Engineer
Objective: To join the staff of an organization that can fully utilize the services of a mechanical engineer in the areas of power and energy systems, energy management, and economic evaluations.
FIELDS OF EXPERIENCE
Industrial Facilities
Public Utilities
Martha Peron
1901 Missouri Avenue
Austin, TX 78702
Home (512) 343-0231 – Work (512) 839-6776
Highlights of Qualifications
• Twelve years experience in manufacturing maintenance with emphasis on electrical, mechanical, hydraulics, pneumatics, drives, overhead, and AC and DC circuits.
• Experienced in planning and coordinating preventive maintenance programs on daily, weekly, monthly, and shutdown basis; includes writing standard operating procedures for PM programs.
• Experienced in warehouse management, ordering, and inventory control.
• Competent, reliable, and committed team player, with ability for increasing levels of responsibility.
Relevant Experience
Administration/Management
• Assisted in planning or coordination of routine shutdown process at a division of a major steel manufacturer.
• Set up and ran an inventory reduction program for a warehouse of a major steel manufacturer; ordered parts for inventory and special projects.

Figure 9-2 Special sections in resumes: the summary or highlights of qualifications sections and the goals and objectives section. Use the highlights section to list key points about yourself. With the highlights section, the potential employer can get a quick picture of who you are professionally. Use the objectives section to indicate your professional focus.

- *Functional approach*—divides your background into areas or groups of related education and experience. For example, you may have taken courses in college on project management, attended professional seminars on the subject, taken lead roles in the management of several projects, and maybe even won an award for your management of a project. All of this could be summarized under the heading "Project Management" in a functionally organized resume (see the schematic illustration of this in Figure 9-1 and a full-content version in Figure 9-2).

The great strength of the functional design is that it consolidates information about your key qualifications, summarizing all relevant work experience and education for each one. Prospective employers looking for someone with project planning and management experience can quickly discern from the functional design whether you have what they are looking for.

Of course, the weakness of the functional design is the strength of the chronological design: In the functional, it may not be immediately clear where, how, and when you gained your experience or education. The chronology of your career is unclear. A solution to this problem is to include a list of your experience and education—no description, just the names and dates (this is schematically illustrated in Figure 9-1 in the heading "WORK HISTORY" and "EDUCATION").

Note If you are at the beginning of your career, or only a few years into it, the functional design may not be a good choice; consider using the chronological design instead.

Highlights Section. Another issue is whether to include a highlights section. This section is popular, particularly for professionals who are several years into their careers. It is particularly helpful in resumes that use the chronological design. In the chronological design, key points about your experience and education are scattered—bits and pieces are cited throughout the work experience and education sections. Readers have to reconstruct your highlights for themselves.

With the highlights section (sometimes called a "summary"), however, you do that reconstruction for your readers. The summary provides a neat bulleted list of your key accomplishments, your key areas of expertise, your key education and training, and so on. The reader might not look further in your resume but would still get a good picture of who you are professionally.

Notice in Figure 9-2 that a bulleted list is used to make the items in the highlights section more scannable. The title of the section can be any variation of things: Summary of Experience, Summary, Highlights of Experience, Summary of Qualifications, Synopsis of Qualifications, Professional Expertise, Qualifications, and so on. You position it just at that point where the eye first makes contact with the page. Many believe that our initial glance makes contact with a page one-fourth to one-third of the way down the page, not at the very top. If you believe that, then putting your very best "stuff" at that point in the resume makes a lot of sense.

Objectives Section. Still another issue in resume design is whether to include an objectives section. The function of this section is to describe your career and professional focus. It can indicate the type of work you want to do, the type of position you aspire to, the type of organization you seek to work for, or some combination of these or other objectives.

This section should be brief—no more than two to three lines. It should also be rather specific and not a patchwork of "sweet nothings." For example, avoid stating your objective as "Seeking a challenging, rewarding career with a dynamic upscale company where I will have ample room for professional and personal growth." (You might as well say you want to be well paid and well fed!) Instead, try for something specific such as the following:

> Construction engineer seeking position in HVAC design and energy calculation for residential and commercial structures.

You may or may not want to include an objectives section in your resume. It's one more thing that defines you—it can pin you down and limit you. However, crafty types rewrite this section to correlate with each new or different position they are seeking. If it's a large, big-city corporation, then words to that effect are in the objectives section. If it's a small, rural company specializing in a particular technology, then words to that effect are in the objectives section. Is this ethical? Perhaps not, except for the very adaptable or the very hungry....

Memberships and Licenses. Another important section in the resumes of many engineers is the section in which they list the professional organizations they belong to and their licenses. In a section like this, list the fact that, for example, you are a member of the American Society of Mechanical Engineers.

Specialized Equipment and Knowledge. Many resume writers want to include a special section on the range of their technical knowledge. For example, computer specialists often want to list all the hardware and software they know well. Electrical engineers might want to indicate that they are skilled in such things as analog circuit and signal analysis as well as digital and control systems.

Miscellaneous Sections. There are many other possibilities for special sections you can include in a resume. For example, if you've published articles in professional journals, create a publications section. If you've received honors and awards, create a section for that. If you have received several patents, then list those in a separate section. If you have various security clearances, list them. The idea is to design the resume so that it emphasizes your best and most important qualifications—special sections with their own headings are a way of doing that.

Personal Sections. Some but certainly not all resume writers include a section at the very end in which they cite loosely relevant personal details about themselves such as interests, nonwork activities, hobbies, memberships, other languages, and so on. Strictly speaking, this sort of information is out of place in the resume—what does the fact that you raise orchids have to do with your career as a structural engineer? Viewed more broadly, however, this kind of information rounds you out as a human being. It gives the prospective employer something to chat with you about while waiting for the elevator—to fill those moments of otherwise uncomfortable silence.

Presentation of Detail. In addition to planning the overall design and contents of your resume, you must also decide on how you want to present the actual details of your background and qualifications.

As Figure 9-3 illustrates, there are many ways to show your experience. You can present it as a simple paragraph, as the lowest of the three examples in the figure shows; you can present it in bulleted lists as the other examples in the figure show. You can highlight the name of the organization you worked for by presenting it first in all caps, bold, italics, or bold italics. Or, you can highlight your title or position by presenting it first, as the rightmost example in Figure 9-3 does.

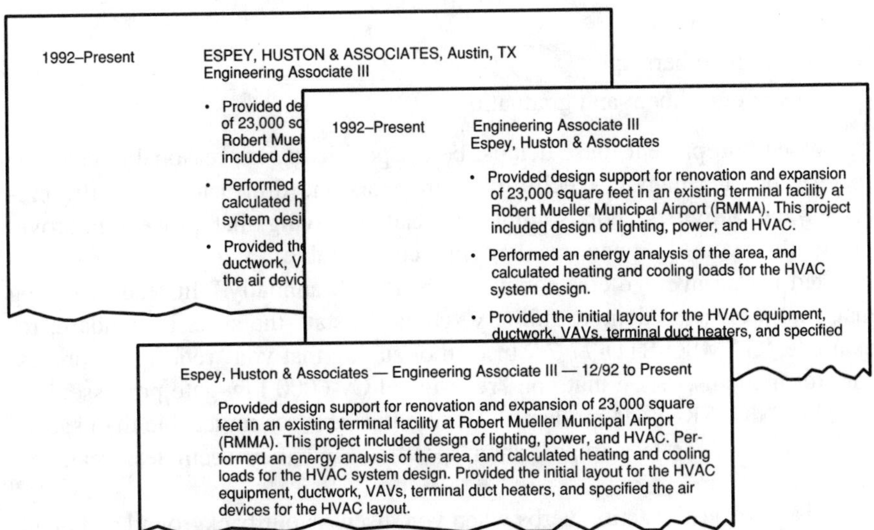

Figure 9-3 Examples of detail formats. Use combinations of list or paragraph format, italics, bold, all caps in the design of the four main elements: date, organization name, job title, and details.

As to the kinds of details you can present in these sections, there are many possibilities, as shown in the following lists. However, remember to be selective; don't include everything in these lists. Don't bury your best qualifications in a mass of less important detail.

For the experience section, consider including these details:

- Name of the organization where you worked, its address, phone number
- Your job title, and your specific responsibilities
- Brief description of the organization, its products, services, technical aspects
- Your major achievements, important projects, promotions, awards
- Experience with technologies, equipment, technical processes
- Dates of employment with the organization.

For the education section, here are some ideas:

- Name of the educational institution, its address
- Brief description of the educational institution
- Your major and minors, grade point average (overall and in your major)
- Major emphasis of study
- Important courses taken
- Brief descriptions of those courses
- Experience with technologies, equipment, technical processes
- Important projects
- Awards, memberships
- Dates of enrollment and graduation

When you present these details, be as specific as you reasonably can: cite specific product names, specific measurements and dimensions, specific processes and activities. For example, instead of saying that process improvements that you designed resulted in "considerable saving," say that they resulted in "an average cost savings of $315,000 annually." Instead of stating that your work was done to military standards, state the specific standard, for example, "SAMSO-STD-77-7." Instead of stating that you "redesigned processors for modems," state that you "redesigned Cy-6000 low-gate processors for QAM/QPSK/FSK-mode modems." Generalities are less noticeable than specifics; they have far less impact than specifics; and they seem less real, less authentic.

Also, use strong action verbs when you discuss your background and qualifications. Verbs like "designed," "developed," "utilized," "coordinated," "supervised," and so on are more striking and memorable than "was involved with," "handled," or "was responsible for."

Format of Headings and Margins. As you design a resume, you'll need to consider the headings and the text in relation to those headings. Many, but certainly not all resumes use a "hanging-head" design in which the headings use the far left margin and the body text of the resume uses an interior margin, indented about 2 inches. This design is particularly effective because it makes the line length of the body text shorter and more easily scannable, the headings more visible, and the sections of the resume more visually distinct.

Resume Length and Headers for Multiple-page Resumes. How long your resume should be depends on how much detail there is in your background and qualifications. It's likely that early in your career you'll have trouble filling up a single page. After a few years, however, it'll be hard keeping your resume to one page, then two pages, and so on. The chief problem with long resumes is that prospective employers may not read them as closely as short ones. If you can somehow cut the length of your resume from three pages to one, the prospective employer is more likely to notice and remember your key qualifications. Some resume experts maintain that you should plan for one page of resume for every ten years of experience. However, there are plenty of reasons why this rule of thumb might not be applicable.

In any case, if your resume is more than one page, place headers at the top of the following pages. Design the header to contain some combination of your name, date, and the page number, as in the examples shown here:

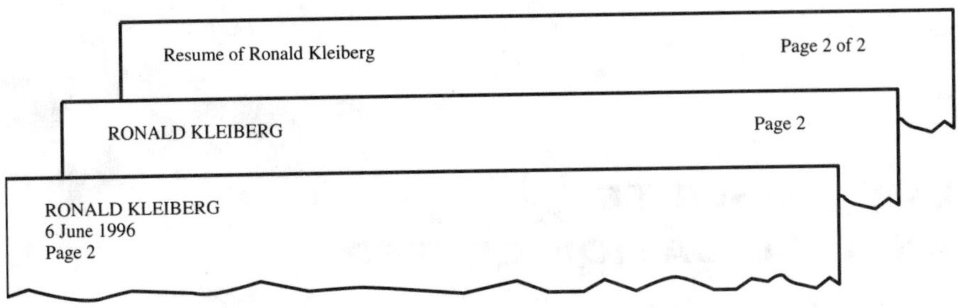

OVERALL FORMAT

Figure 9-4 gives you a schematic view of some common ways to design the overall format of resumes. You can see that headings can be centered, they can be placed on the left margin but run into the text, or they can be put in their own column separate from the text. Some resumes add ruled lines horizontally or even vertically to increase the visual separation of the components of a resume. (See Figure 9-5 for a full-content example of an engineering resume.)

Resume headings: titles such as "Experience," "Education," and so on.
Centered headings and headings flush with regular text.
Resume text (background, details, description, etc.).
"Hanging-head" format: commonly used in resumes. It shortens the length of the text lines, makes the headings more prominent, and makes the sections of the resume more separate and distinct.

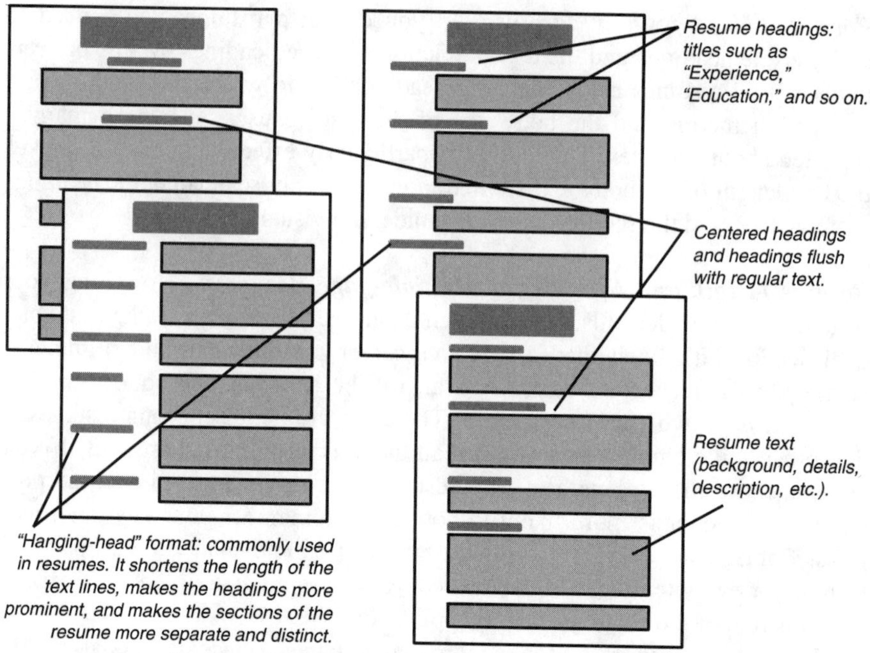

Figure 9-4 Various possibilities for resume design. Think about the overall design of your resume first—how the headings are positioned in relation to the text; visualize it in blocks like these, without the words, to get an overall sense of the design.

HOW TO WRITE AN APPLICATION LETTER

Usually accompanying the resume is an application letter, sometimes called a cover letter. This letter is the first thing that potential employers see when they unfold the contents of the envelope—the application letter on top, with the attached resume beneath it.

As mentioned earlier, whether to include an application letter with your resume depends on the potential employer and the advertisement. Sometimes, only the resume is requested; in a few instances, only the letter. Sometimes, after prospective employers make their initial selection of candidates, they request the other of the two components.

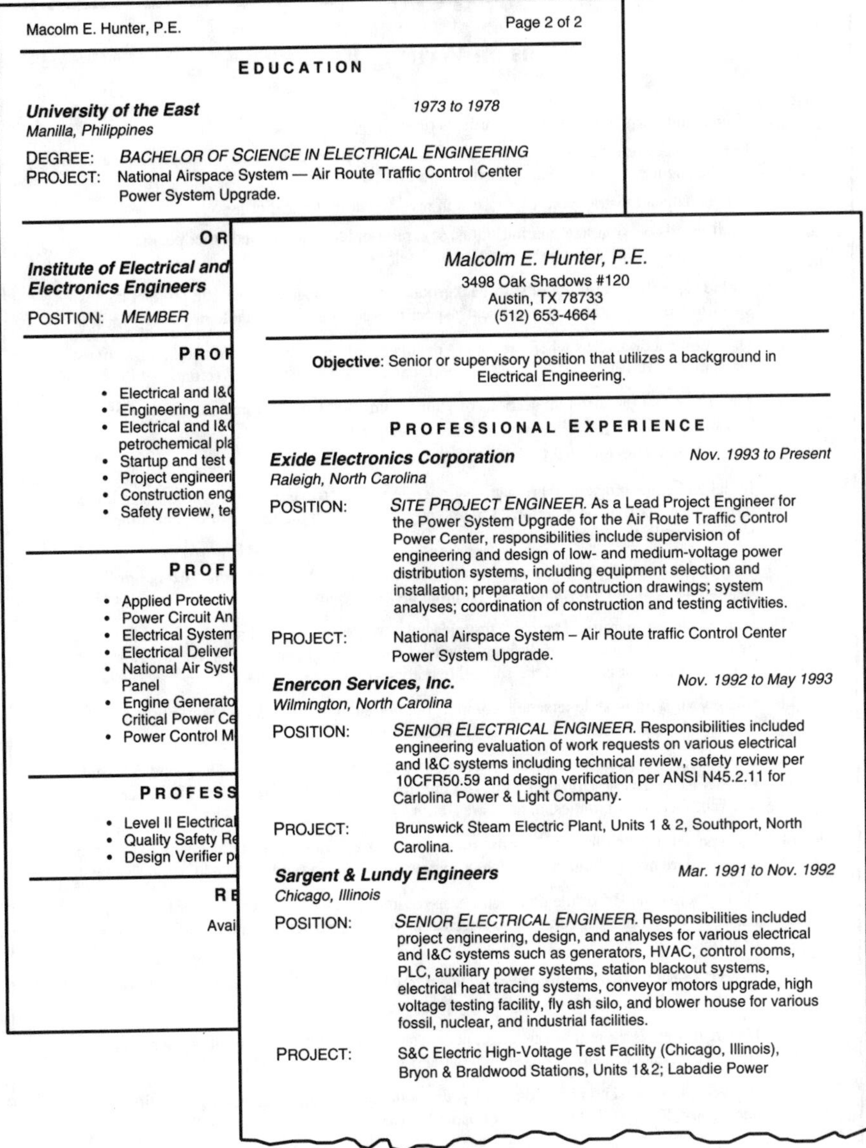
Macolm E. Hunter, P.E. Page 2 of 2

EDUCATION

University of the East *1973 to 1978*
Manilla, Philippines

DEGREE: *BACHELOR OF SCIENCE IN ELECTRICAL ENGINEERING*
PROJECT: National Airspace System — Air Route Traffic Control Center Power System Upgrade.

OR

Institute of Electrical and
Electronics Engineers
POSITION: *MEMBER*

PROF

- Electrical and I&C
- Engineering anal
- Electrical and I&C petrochemical pla
- Startup and test
- Project engineeri
- Construction eng
- Safety review, te

PROF

- Applied Protectiv
- Power Circuit An
- Electrical System
- Electrical Deliver
- National Air Syst Panel
- Engine Generato Critical Power Ce
- Power Control M

PROFESS

- Level II Electrical
- Quality Safety R
- Design Verifier p

RE

Avai

Malcolm E. Hunter, P.E.
3498 Oak Shadows #120
Austin, TX 78733
(512) 653-4664

Objective: Senior or supervisory position that utilizes a background in Electrical Engineering.

PROFESSIONAL EXPERIENCE

Exide Electronics Corporation *Nov. 1993 to Present*
Raleigh, North Carolina

POSITION: *SITE PROJECT ENGINEER.* As a Lead Project Engineer for the Power System Upgrade for the Air Route Traffic Control Power Center, responsibilities include supervision of engineering and design of low- and medium-voltage power distribution systems, including equipment selection and installation; preparation of contruction drawings; system analyses; coordination of construction and testing activities.

PROJECT: National Airspace System – Air Route traffic Control Center Power System Upgrade.

Enercon Services, Inc. *Nov. 1992 to May 1993*
Wilmington, North Carolina

POSITION: *SENIOR ELECTRICAL ENGINEER.* Responsibilities included engineering evaluation of work requests on various electrical and I&C systems including technical review, safety review per 10CFR50.59 and design verification per ANSI N45.2.11 for Carlolina Power & Light Company.

PROJECT: Brunswick Steam Electric Plant, Units 1 & 2, Southport, North Carolina.

Sargent & Lundy Engineers *Mar. 1991 to Nov. 1992*
Chicago, Illinois

POSITION: *SENIOR ELECTRICAL ENGINEER.* Responsibilities included project engineering, design, and analyses for various electrical and I&C systems such as generators, HVAC, control rooms, PLC, auxiliary power systems, station blackout systems, electrical heat tracing systems, conveyor motors upgrade, high voltage testing facility, fly ash silo, and blower house for various fossil, nuclear, and industrial facilities.

PROJECT: S&C Electric High-Voltage Test Facility (Chicago, Illinois), Bryon & Braldwood Stations, Units 1&2; Labadie Power

Figure 9-5 Excerpts from the resume of an experienced professional engineer. Notice the use of small caps for position titles (such as "Site Project Engineer"). The headings on page 2 of this resume are "Education," "Organizations," "Proficiencies," "Professional Training," "Professional Certification," and "References."

Tips on Writing Resumes

1. Don't omit normal words such as articles (*a, an, the*).
2. For sentences referring to your own work, omit "I." Instead of writing "I supervised a team of 12 designers..." write "Supervised a team of 12 designers..."
3. Present education and work experience in reverse chronological order.
4. Omit details on your age, marital status, sex, religion, handicaps, and other personal matters. Don't include a photograph of yourself.
5. Include specific details about your background and qualifications: specific product names, specific dimensions and measurements, specific processes and technologies.
6. Use strong action verbs when presenting details about your background and qualifications: "designed," "developed," "utilized," "coordinated," "supervised," and so on.
7. Make sure that the different sections of your resume are distinct from each other; use spacing, ruled lines, and headings to indicate their boundaries.
8. Avoid lengthy paragraphs; keep paragraphs under 6 or 7 lines.
9. Indicate the meanings of abbreviations or acronyms—don't assume the whole world knows what "GPA" or the construction "3.5/4.0" means. Spell out the names of organizations; briefly explain their functions.
10. Use format consistently: if you present the details of your work experience using one format, use that same format in other similar areas of your resumes.
11. Use consistent margins. Typically resumes have two left margins, for example, in the hanging-head format, a far-left margin where headings align, and at least one interior margin where text aligns. Make sure all text aligns to these margins.
12. Make your writing style terse and compact but not unintelligible. Don't omit normal articles. Do, however, omit "I" when discussing yourself.
13. Use special typography moderately: use bold, italics, underlining, type sizes, and different fonts conservatively. Don't go wild with multiple fonts (Times, Helvetica, Thames, etc.) and font faces (bold, italics, underscores, etc.).
14. Use special typography consistently: for example, if you put company names in bold in the work-experience section, put college or university names in bold in the education section.
15. While there is no fixed rule on length, keep resumes as short as possible—for example, one page at the start of your career; two pages after you've gained substantial professional work experience.
16. Keep the resume from spilling over by just a few lines to a second or third page. Force the resume to fill the pages it occupies.
17. If your resume is more than one page, put a header on the second and following pages. Design the header like the ones shown in the preceding pages.
18. It is acceptable to send photocopies of your resume, but if you do, get a high-quality photocopy. Request that high-quality paper be used.

There are two broad categories of letters, based on the information they contain:

- *Cover letters*—In the true cover letter, you do nothing more than announce that a resume is attached and that the purpose is to investigate an employment opportunity with the addressee. This type of letter specifies the position

1307 Marshall Lane
Austin, TX 78703
25 May 1996

Ms. Juanita Jones
Hughes & Gano, Inc.
P.O. Box 1113
Austin, TX 78758

Dear Ms. Jones:

Please accept the attached resume as application for the position of Process Engineer currently available with your company.

I'll be looking forward to meeting with you at your earliest convenience. I can be reached at (512) 471-4991 during regular working hours or at (512) 471-8691 in the evenings and weekends.

Please contact me if you need any further information about my background or qualifications.

Sincerely,

Patrick H McMurrey

Patrick H. McMurrey
Encl.: resume

Figure 9-6 Cover letter: a brief correspondence that identifies the position being sought and the purpose of the correspondence. For most job searches, use the fully detailed application letter, as described in this chapter.

you are seeking. As illustrated in Figure 9-6, this is a very brief letter, the body text totaling less than ten lines. If the job advertisement asks for resumes only, you can include this type of letter to identify the position you're inquiring about.

- *Full application letters*—In the true application letter, you discuss your background and qualifications as relevant to the position you are seeking. The job of this letter is to promote yourself—to highlight the reasons why you are right for the position. This type of letter is the focus of the rest of this chapter.

Which to use? The cover letter is certainly easier to write, but it doesn't do anything for you. The full application letter acts as your proxy, showing the prospective employer specifically which of your qualifications makes you right for the job. And, in cases where a full application letter is expected, a minimal cover letter may make you sound noncommittal or even indifferent.

CONTENTS AND ORGANIZATION OF THE APPLICATION LETTER

The function of the application letter is to introduce you to the prospective employer, state the purpose of the letter (which is to seek an interview), identify the position you're inquiring about, and summarize your relevant background and qualifications. This last function is the most important. The application letter is not just another form of the resume—it is a very selective view of it. It makes a direct and overt case for you as a good candidate for the employment. It does this primarily by pointing out aspects of your background that are a good fit with the specific job you are seeking.

The following discusses common contents and organization used in application letters. Remember that this is only a review of what's typically done, not what is the "right" way. You may need to take a different approach because of your background, your qualifications, or the job you are seeking.

First Paragraph. The first paragraph of the letter should be brief and do some combination of several specific things: state the purpose of the letter (to inquire about employment); state how you found out about the opening, if applicable; make one statement that will catch readers' attention and make them want to continue reading. And that's it. Keep this first paragraph short, five to six lines at the maximum.

In this first paragraph, consider using one of several common tricks to catch your readers' attention:

- State one important fact about your background or qualifications that makes you the right candidate for the position.
- Cite some bit of information about the company you are applying to—knowledge that shows you are informed and that relates to the position you are seeking.
- If possible, mention the name of someone in the company who knows you and can speak knowledgeably—or, better yet, favorably—about you.
- Make some sort of enthusiastic, energetic statement about the kind of work you want to be doing, the kind of organization you want to be working for, or maybe something about your long-range professional goals and objectives. If you are writing the letter "cold" (not in response to any job advertisement), say something about why you want to work for the company—

and take this opportunity to show that you know something specific about the company.

Whichever of these strategies you use, remember to keep it short (for example, you probably can't combine several of these and stay under the five- or six-line maximum). Also, remember to write this part of the letter, as well as the rest, in terms of the potential employers' perspective, and only minimally, yours. For example, employers may not want to hear at length about all your personal goals—only enough to see that you fit in with their operations.

Middle Paragraphs. The middle portion of the letter should discuss those of your qualifications that are important in their own right and that relate specifically to the employment you are inquiring about. Usually, somewhere in these paragraphs, you suggest that readers can see the attached resume for more detail. For these paragraphs, you can use the same kinds of organization as in the resume:

- *Chronological approach*—In the chronological approach, you discuss your educational background in one or more paragraphs, and then your work experience in another set of paragraphs. Put work experience before education if it is your best "stuff." (And of course if your educational background is "old history," just leave it out.)
- *Functional approach*—In the functional approach, you focus on the important *areas* in your background and qualifications—for example, project management, research and development, quality control, vendor coordination, documentation, and so on. Ideally, you'd have a separate paragraph for each of these areas of your experience. In each of these paragraphs, you'd discuss anything in your background, whether work experience or education, that relates to that area.

If you find that your discussion of your background and qualifications just takes too many lines, no matter how much you cut or condense it, consider using a bulleted list (as illustrated in Figure 9-9). This format enables you to present all your important details but in a more readable and scannable way.

Final paragraphs. In the final portion of the letter, you wrap it up: mention that the resume is enclosed if you've not already done so; urge the prospective employer to get in touch; indicate times you are available to interview and just generally facilitate the arrangement of an interview; and find some parting encouraging, enthusiastic comment to make, such as your strong interest in the employment, the profession, the company, and so on. Some writers indicate that they will call the prospective employer on a certain date (for example, a week after they have mailed the letter). While others might find this tactic too aggressive, it certainly puts pressure on the employer to take action.

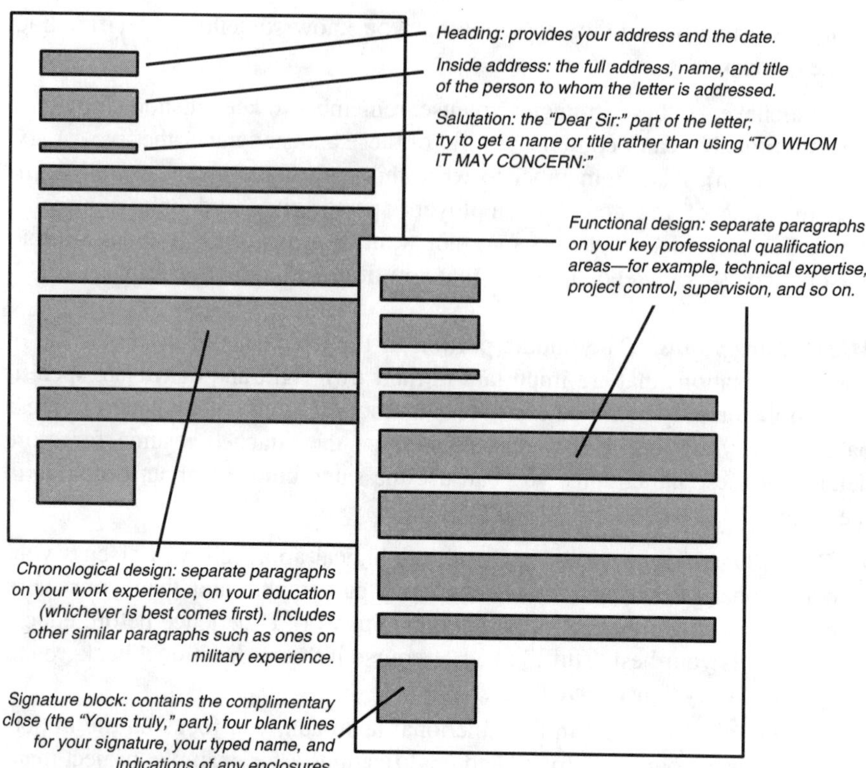

Heading: provides your address and the date.
Inside address: the full address, name, and title of the person to whom the letter is addressed.
Salutation: the "Dear Sir:" part of the letter; try to get a name or title rather than using "TO WHOM IT MAY CONCERN:"
Functional design: separate paragraphs on your key professional qualification areas—for example, technical expertise, project control, supervision, and so on.
Chronological design: separate paragraphs on your work experience, on your education (whichever is best comes first). Includes other similar paragraphs such as ones on military experience.
Signature block: contains the complimentary close (the "Yours truly," part), four blank lines for your signature, your typed name, and indications of any enclosures.

Figure 9-7 Common sections of application letters. You can organize the letter chronologically or functionally the same as you can the resume.

FORMAT OF APPLICATION LETTERS

Consult the following suggestions for questions about the format of the application letter:

- Use a standard business-letter format, such as the one shown in the examples in this chapter. (See Ch. 4 for style and format of business letters in general.)
- Singlespace the individual components (never doublespace). Doublespace between the components.
- Leave four blank lines between the complimentary close and your typed name, and sign your name in that space.
- Do not indent the first line of paragraphs of the body text. Use standard left and right margins; 1 inch, 1.5 inches, or 2 inches are all acceptable. Use wider margins when you have trouble filling up the page.

- Use standard top and bottom margins: The letter can begin anywhere from 1 inch to 3 inches from the top edge of the page; it should end no closer than 1 inch to the bottom edge. Use the variability of these dimensions to fill up the page.
- Carefully position your letter on the page. If your letter is short, don't leave it crammed up at the top of the page. Use the variables of margins and spacing between text components to position the text of the letter in the upper middle of the page.
- Avoid dense paragraphs. Don't expect readers to labor through paragraphs over ten or twelve lines long. Use paragraph breaks and numbered or bulleted lists to loosen up dense information in the middle paragraphs.
- For additional eye appeal, consider creating an attractive, professional-looking design for your name and address, such as the ones illustrated in Figure 9-8.

Tone in Application Letters. In an application letter tone may be one of the most important of all aspects, but also the hardest to define. Tone should reflect your view of yourself and the type of professional you want to be. However, you don't want your letter to sound brash, arrogant, or overconfident if that is not the personality you'd like to project. Application letters can develop bad tone as the result of good intentions:

- *Stiff, overly formal, overly reserved.* When you write an application letter, the pressure is on—obviously. One tendency is to freeze up and be overly cautious. Ironically, this can sound like nonchalance or indifference. And, in any case, the stiff, reticent, overly formal tone may suggest a kind of person who prospective employers won't want to work with.
- *Intimidatingly qualified, or even overqualified.* Tone in an application letter can also go bad when you overemphasize your qualifications, making yourself sound like a superhero. Readers are likely to get uneasy with candidates like this—they fear the prospect of working with an overdeveloped ego; they worry about the safety of their own jobs; they wonder how "legit" you are, whether you're stretching the truth or simply lying.
- *Unctuous, fawning, flattering.* It's possible to try too hard to sound bright, positive, enthusiastic, eager; it's possible to sound phony in saying nice things about the prospective employer.
- *Hungry, desperate.* Avoid the tone that says "I'll do anything!" The ready, eager, and willing tone can go bad when it starts sounding desperate for a job—any job. In your job search maintain your professional focus and integrity—you won't do just any kind of work; you want the kind of work you have trained for.

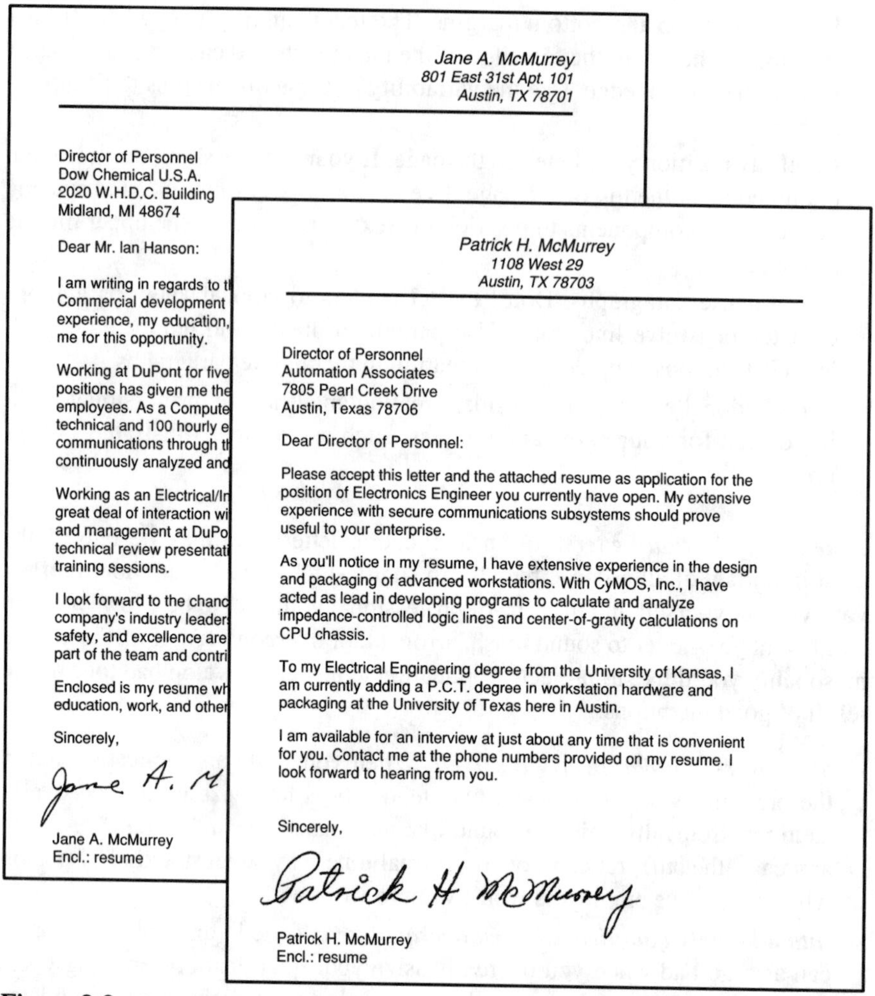

Jane A. McMurrey
801 East 31st Apt. 101
Austin, TX 78701

Director of Personnel
Dow Chemical U.S.A.
2020 W.H.D.C. Building
Midland, MI 48674

Dear Mr. Ian Hanson:

I am writing in regards to t
Commercial development
experience, my education,
me for this opportunity.

Working at DuPont for five
positions has given me the
employees. As a Compute
technical and 100 hourly e
communications through t
continuously analyzed and

Working as an Electrical/In
great deal of interaction wi
and management at DuPo
technical review presentati
training sessions.

I look forward to the chanc
company's industry leader
safety, and excellence are
part of the team and contri

Enclosed is my resume wh
education, work, and other

Sincerely,

Jane A. M

Jane A. McMurrey
Encl.: resume

Patrick H. McMurrey
1108 West 29
Austin, TX 78703

Director of Personnel
Automation Associates
7805 Pearl Creek Drive
Austin, Texas 78706

Dear Director of Personnel:

Please accept this letter and the attached resume as application for the position of Electronics Engineer you currently have open. My extensive experience with secure communications subsystems should prove useful to your enterprise.

As you'll notice in my resume, I have extensive experience in the design and packaging of advanced workstations. With CyMOS, Inc., I have acted as lead in developing programs to calculate and analyze impedance-controlled logic lines and center-of-gravity calculations on CPU chassis.

To my Electrical Engineering degree from the University of Kansas, I am currently adding a P.C.T. degree in workstation hardware and packaging at the University of Texas here in Austin.

I am available for an interview at just about any time that is convenient for you. Contact me at the phone numbers provided on my resume. I look forward to hearing from you.

Sincerely,

Patrick H McMurrey

Patrick H. McMurrey
Encl.: resume

Figure 9-8 Examples of application letters: Notice that the first paragraph of the letter on the right identifies the position being sought and makes one strong, general statement about the writer's qualifications. (The fancy headings are not a requirement—just a nice, eye-catching touch.)

- *Overly humble, overly simple, above it all.* What's wrong with adopting an attitude that says "this is who I am, this is what I can do, this is what I have done—take it or leave it"? It's simple, humble, plain, no nonsense. That approach can run aground as well. It can sound so excessively humble that you project yourself as superior, above-it-all, even arrogant.

Patrick H. McMurrey
1108 West 29
Austin, TX 78703
(512) 471-9229 (home) (512) 878-6556 (work)

Director of Personnel
Automation Associates
7805 Pearl Creek Drive
Austin, Texas 78706

Dear Director of Personnel:

Attached is a copy of my current resume. I would appreciate your time in evaluating my qualifications in relation to your current needs for a Senior Electrical Engineer in TDB's large building design projects.

I have over 15 years experience in various facets of electrical design and engineering. Also, I have experience in power and control design including analyses for power generation; low-, medium-, and high-voltage power distribution systems; fire detection and protection systems; plant security systems, programmable logic controllers (PLSc); as well as equipment layout for various types of industrial facilities.

I am currently employed with Exide Electronics Corporation as Site Project Engineer in the National Air-Space Federal Systems Engineering Division. My current responsibilities are as follows:

- Lead project engineer for the power system upgrade of Denver, Albuquerque, Indianapolis, and Jacksonville Air Route Traffic Control Centers (ARTCC). This power upgrade is part of Exide's current contract with the U.S. Air Force and the Federal Aviation Administration.

- Supervision of varying numbers of electrical engineers and designers in various general engineering tasks for projects such as low- and medium-voltage power distribution system engineering and design.

- Vendor interface for installation of equipment such as diesel generators, switchgears, and power control monitoring systems.

This current work and past projects, along with the references I am including in this letter, all attest to my solid record of initiative, responsibility, creativity, and professional dedication. I am an effective, contributing member of any organization that I am associated with. If you are interested in discussing my experience and capabilities further, please contact me at one of the numbers shown above.

Sincerely,

Patrick H McMurrey

Patrick H. McMurrey
Encl.: resume, reference list

Figure 9-9 Example of an application letter: Notice how much specific detail the writer packs in concerning his experience. Notice also how the bulleted list relieves some of the density of the letter.

Bad tone can start from good intentions: We want to be cautious and respectful; to show what's good about ourselves; to be enthusiastic and complimentary; to sound comfortable and confident professionally; to demonstrate that we are intense and earnest about the employment opportunity; and to be honest and straightforward. But if any of these get overemphasized or expressed in the

Tips on Writing Application Letters

1. Avoid diving headlong into the details of your background and qualifications from the very first paragraph. Create an introductory paragraph that performs the functions mentioned earlier in this chapter.
2. Try to get a specific name or department to which to address the letter; avoid the "TO WHOM IT MAY CONCERN" syndrome.
3. Make the letter individualized to the addressee. Even if you are in a massive job search and are sending out many letters, avoid sounding as though you're a zombie writing form letters; make them sound as though they were written uniquely for the specific recipient.
4. Be sure to mention that your resume is enclosed with the letter.
5. Use standard business letter format in the application letter, as shown in the examples in this chapter and as described in detail in Chapter 4. (And remember to punctuate the salutation with a colon, not a comma!)
6. Keep the paragraphs of the letter short: keep the introductory paragraph particularly short, under 5 to 6 lines; keep the background paragraphs under 10 to 12 lines.
7. Seek a nice, bright, energetic, positive tone. Watch out for the problems with tone discussed in this chapter. (Get someone to read a rough draft of your letter and describe the kind of personality it projects.)
8. Write the letter in terms of the prospective employer's needs or interests, and only minimally your own. Discuss yourself in terms of what the prospective employer needs.
9. Use the full application letter (as opposed to the minimal cover letter which gives no details on your qualifications and background) unless the job advertisement specifically requests only the resume.
10. Avoid negative discussion of previous employers; generally avoid stating reasons why you left previous jobs.
11. Unless specifically requested by the prospective employer, avoid discussion of salary, benefits, or other compensation.
12. Keep the letter to one page.
13. Don't state anything new in the application letter—you're elaborating upon qualifications you've already listed in the resume.
14. While it's acceptable to send out high-quality photocopies of the resume, the letter should be freshly typed or printed out. The letter needs to give the appearance of being specially prepared for the addressee.
15. Avoid spelling, grammar, usage errors, and bad writing at all costs!

wrong way, problems of tone occur and we run the risk of projecting the wrong image of ourselves.

HOW TO WRITE A FOLLOW-UP LETTER

Write a follow-up letter when you've not heard from the prospective employer after two weeks, after you've had an interview, when you want to acknowledge a refusal of a job offer, and when you must reject or accept a job offer. The most important use of the follow-up letter is for those situations when you are waiting (and waiting) and have had no word from the prospective employer. (See Figure 9-10 for an example of this type of follow-up letter.) To write this type of follow-up letter, consider including these contents:

- State the reason you are writing the letter—to inquire about the application letter and resume you recently sent.
- Indicate the date you sent the letter and the resume, and specify the position you were inquiring about.
- Suggest that the letter and the resume might have been lost in the mail or routed incorrectly somewhere within the recipient's organization.

801 East 31st Street #101
Austin, Texas 78701
3 March 1996

Director of Personnel
Automation Associates
7805 Pearl Creek Drive
Austin, Texas 78706

Dear Director of Personnel:

On February 17, I applied for a position as manufacturing engineer with your firm. Not having heard from you in the two weeks since that time, I'm concerned that my letter may have been lost.

Attached is a copy of the original letter and resume that I sent. As you will see, they detail my work experience, my education, and my sincere interest in working for your company.

If you have made a decision about this position, I would appreciate hearing from you. My availability and my interest continue, and I look forward to the chance to discuss the job and my background with you in person.

Sincerely,

Jane A. McMurrey

Jane A. McMurrey
Encl.: Copy of 2-17 letter and resume

Figure 9-10 Follow-up to an application letter: Although the follow-up letter can be used for different reasons, its most important use is to inquire about the fate of an application letter and resume that you have sent but have received no response.

- Enclose a copy of the original letter and resume, and state in the follow-up letter that you have enclosed them.
- As tactfully as you can, encourage the recipient to let you know of any decisions that have been made about the job (perhaps indicating that your own decisions are dependent upon it).

EXERCISES

Interview at least three professional engineers concerning the application letters and resumes they typically see when hiring new engineers. Ask them questions like the following:

1. How do they "read" resumes: line by line from beginning to end? If they skip around and scan, what do they look for? What catches their eye? How important are specific details such as brand names, model numbers, titles of specifications, and dimensions in resumes?
2. What can the engineer who is just graduating and getting started in the profession legitimately put in the work-experience section of a resume?
3. Should personal information such as hobbies, community activities, or reading interests be kept out of resumes? If not, what purpose do they serve?
4. What are the typical problems that cause a resume to be skipped or scarcely glanced at? How much does the formatting of a resume contribute to their willingness to read a resume carefully and thoroughly? What effect does heavy use of bold, italics, all caps, and different font and font sizes have on the way they read a resume?
5. Are applicants asked to send only a resume or only an application letter? Do they expect to see a simple cover letter (as described in this chapter), or do they expect a full application letter highlighting the applicant's relevant and important qualifications?
6. In their view, what is the chief function of the application letter?
7. Does tone ever cause a problem in these letters? Do details and specifics in an application letter matter, or should the letter be general?

Once you've collected this information, summarize it in a memorandum report addressed to your instructor, using the format shown in Chapter 4.

BIBLIOGRAPHY

The following books and articles provide more information on the topics covered in this chapter:

Anton, Jane L., Michael L. Russell, and the Research Committee of the Western College Placement Association. *Employer Attitudes and Opinions Regarding Potential College Graduate Employees.* Hayward, CA: Western College Placement Association, 1974.

Beatty, Richard H. *The Resume Kit.* New York: Wiley, 1991.

Bostwick, Burdette F. *Resume Writing: A Comprehensive How-To-Do-It Guide.* New York: Wiley, 1985.

Jackson, Tom. "10 Musts for a Powerful Resume," in *CPC Annual,* 30th ed. Bethlehem, PA: College Placement Council, 1986.

Kennedy, J., and T. Morrow. *Electronic Resume Revolution.* New York: Wiley, 1995.

Lewis, Adele. *The Best Resumes for Scientists and Engineers.* New York: Wiley, 1988.

U.S. Department of Labor. *Job Search Guide: Strategies for Professionals.* Washington, D.C.: GPO, 1993.

U.S. Department of Labor. *Tips for Finding the Right Job.* Washington, D.C.: GPO, 1992.

10

WRITING WITH COMPUTERS

The main focus of this chapter is writing with computers. Included here are reasons for using computers to do your writing, some tips on features to look for in writing software (our term for word-processing software), and tips on writing with computers. In addition, we focus briefly on some useful tools you'll probably need in addition to your writing software, such as software for graphics, scanning, virus protection, and file compression, as well as scanning hardware. (Brief discussions of e-mail and the Internet can be found in Chapter 4, "Writing Letters, Memoranda, and Electronic Mail.")

SOFTWARE FOR WRITING

Probably one of the most important ways you will use computers as an engineer is for written work such as business letters, memoranda, reports, articles, and proposals.

WHY WRITE WITH A COMPUTER?

There are not too many of this type left any more, but a certain segment of professionals refuses to use computers for any of their tasks, or, if they do use computers, they do so grudgingly. If you happen to be one of these antediluvians, consider these reasons for using computers in your professional writing tasks:

- *Productivity*—You scribble it out on paper and then have to transcribe it into a computer file or into a typewritten document—that's duplicate labor and extra time.
- *Ease of revision and update*—If you find problems that require correction or rewriting, you may have to retype the whole thing, but with a computer you just edit the file and reprint. Writing software makes it easy to copy, delete, or move whole paragraphs.
- *Comfort*—With computers, you trade writer's cramp for carpal tunnel syndrome—no, seriously, writers who get used to the computer keyboard typically complain that it's actually painful to go back to a pencil or pen for long periods of writing.
- *Adaptability*—Once your text is in an electronic file, you can do all kinds of things with it—copy it into another document, send it to a colleague over the Internet, use it as a template for other similar documents, and of course continue to update it.
- *Legibility*—Even with typos, your printout file will be more legible than the handwritten version. Your colleagues will appreciate it!
- *High-quality, professional-looking output*—With advanced word-processing software, and certainly with desktop publishing software, you can produce documents that look like professionally designed and printed material. This saves you time and money, not to mention strengthening your professional image.
- *Lack of secretarial support*—Since the 1980s, it has become the norm in many organizations that almost no secretarial support is available. That means *you* type your own letters, memos, reports, articles, and so on. It has been a long time since any but the highest executives could scribble out notes on a legal pad or mutter something into a dictation machine and toss it to a secretary for transcription.
- *Speed*—Most writers who use computers feel that they can produce text faster on the computer keyboard than with pencil and paper or with a typewriter. And of course, the rest of the documentation process can be greatly accelerated by computers.
- *Data sharing*—Many of your professional projects that involve writing are likely to involve collaboration—team-writing as well as other sorts of teamwork. To be a fully contributing member of the team, you need to work with the same tools as the rest.

WHICH WRITING SOFTWARE TO USE?

You have a wide range of writing software to choose from. In making your choice, consider your needs, your equipment, and the environment you work in.

If you're new to writing with computers, the terms used to refer to writing software may seem confusing.

- *Text editors.* The simplest writing software comes with your operating system. OS/2 provides the E editor; Windows, the Notepad and the Write programs; UNIX systems provide the vi editor. Plus, there are many shareware text editors, such as GNU Emacs, available on the Internet. These tools give basic text-entry and text-revision capabilities plus some word-processing conveniences such as global search and replace, margin setting, spell checking, and some macro ability. This level of writing software is very limited, however. You get few if any of the tools necessary for producing professional-looking documents.
- *Advanced word-processing software.* While there are no clear dividing lines in this area, "advanced" writing software enables you to produce documents that look professionally printed. You can change type styles (fonts), type sizes, type faces (bold and italics); add certain kinds of graphics; and "automate" page and figure-title numbering as well as cross-referencing.
- *Desktop publishing software.* Distinctions between advanced writing software and desktop publishing software are difficult. Many of the features considered to be the province of desktop publishing systems have become available on advanced writing software such as WordPerfect, Word, and AmiPro. What's the difference? Desktop publishing software is more "robust"—it's built to handle bigger jobs, handle more exceptions, and provide far greater flexibility in general. For example, if you are producing a library of a dozen books, some of them over 500 pages, desktop publishing software may be your only choice. Desktop software, such as Interleaf, Pagemaker, Ventura, FrameMaker, and QuarkXpress, requires more powerful computer hardware than most of us (well, some of us) can afford; plus the learning curve is considerably steeper.

One of the most important considerations in choosing writing software is what your colleagues or clients are using or what they expect. It's difficult at best to convert files written with one type of software to another type of software—sometimes you lose all of your careful formatting (bold, italics, special margins, bulleted lists, and so on). Another consideration is the level and complexity of the work you're doing—if it's relatively brief proposals and reports, advanced writing software is more than adequate.

And of course there is much to be said for simple text editors—they don't distract you with all those fancy formatting choices. They clear the decks, so to speak, so that you can concentrate on writing. And once you use simple text editors to get your words right, you can pull the text into the sophisticated writing software required by your clients and do the fancy formatting there.

Can Your Word-Processing Software Do This?

When you're evaluating writing software, check for features like these as well as their ease of use. And, finally, once you get your writing software, take some time to get acquainted with each of these features.

- *Standard word-processing features*—Except for the most primitive word-processing programs, you should expect such functions as search and replace (global or local); margin and tab controls; and block copying, moving, and deleting.
- *Automated numbering*—Check how the software handles page numbering. Can you suppress numbers on certain pages, choose where to place the numbers, specify the number at which to start, use lowercase Roman numerals for front matter and then switch to Arabic numerals within the same file? Does the software automate numbered lists? Can it automate the numbering of chapters, figures, and tables?
- *Graphics*—Check to see what sorts of graphic tools the software has. Ideally, it should have a robust set of tools for creating graphics *within* a document as well as plenty of flexibility for importing graphics from other applications (graphics software such as AutoCAD or CorelDraw, different graphics formats, as well as graphics created in other word-processing programs).
- *Tables*—How writing software handles tables is often the acid test. Check to see how much control you have over fonts, alignment, and cell size. Check to see whether you can convert columns of regular text to tables, and check to see if you can "dump" a table back into columns of text.
- *Formulas, equations, and special characters*—Look carefully for how well the software handles special characters, formulas, and equations. As you probably know, equations make heavy use of special formatting—different type sizes, special vertical and horizontal alignment, as well as plenty of special characters.
- *Footers and headers*—Check to see whether the writing software you're investigating can insert running headers and footers into your documents. Do you get full control over the style and placement of those headers and footers? Can you change the contents or style of footers and headers *within* an individual document?
- *Fonts and faces*—Most of the leading software packages offer plenty of fonts, but check to see how much flexibility you have when using fonts and whether a reasonable range of sizes (for example, 48 to 6 points) and faces (such as bold, italics, underscore, strikethrough, superscript, subscript, reverse video, shadowed) is available for each font.

Can Your Word-Processing Software Do This?

(continued)

- *Spell checking and thesaurus*—Check on the size (number of words) of the dictionary, and find out whether you can create or acquire specialized dictionaries—for example, one for a particular engineering field.
- *Indexing*—Does the software support automated indexing (in which you insert index tags, normally hidden in the document)? How many levels of index entries?
- *Table of contents*—Can you generate tables of contents automatically? Can you control which headings appear in the TOC?
- *Landscape and portrait modes*—Portrait mode is the standard orientation for pages like the ones you are looking at. However, you may need the sideways orientation, called "landscape," for various projects. For example, you may have a long table or organizational chart that won't fit on a portrait-mode page. Check to see whether the software will permit changing between these modes within the same document.
- *Multiple-column text*—Advanced writing software ought to handle two or more columns of text. Check to see whether the software can handle changes in columns *within* a file or document. (Some software forces you to stay with one columnar setting for an entire file.) Check also to see how much control you get over column width, balance, and pagination.
- *In-document calculations*—Some writing software enables you to perform or automate calculations within documents. For example, you can arrange for columns of numbers in a table to be totaled or averaged automatically; you can do various calculations on numbers that appear in running text.
- *Macros and keystroke replay*—Check out the macro capabilities of the writing software. These as well as styles, templates, and libraries (explained in the following) are critical time-savers. Imagine that in a big report the name of the technology you're discussing is terribly long. Put it in a macro that will zap it in by merely pressing two keys!
- *File management*—Your writing software should enable you to display and change directories, select files to edit from directories, as well as copy, delete, rename, and move files—all *without* having to leave the writing software.
- *Importing or linking to external data*—Currently, a hot feature in advanced writing software is the ability to access data, even dynamically, in other applications. For example, you can import data from a Lotus spreadsheet and have it appear right there in your text—without

Can Your Word-Processing Software Do This? *(continued)*

having to open the spreadsheet, copy the data, then paste it into your document. And if the data is linked dynamically, it changes when the data in the spreadsheet changes.

- *Screen capture*—Not many writing software programs do, but check to see if the software can actually "capture" (make a copy) of some part or all of what's displayed on the screen. (Typically, you paste it into a graphic frame once you've captured it.)
- *Revision indicators and version control*—Look for a feature that enables you to mark new or changed text in a document (often these show up as "change bars," thick vertical lines in the margin to the side of new or changed text). A related feature involves the ability to hide or show certain portions of text. Imagine a document is to be used with two different clients: version control would enable you to hide the material for client A and show the material for client B.
- *Merge function for form files and data files*—Check to see whether the software you're considering can merge data files and text files. The typical use of this feature is mass mailing: you have a data file with names and addresses and a form letter. Instead of copying the letter a zillion times and changing the name and address on each one individually, you can set up a data file and a form file and let the software "merge" them. Your zillion personalized letters can be printed out or saved to your hard disk. Also, check to see whether the writing software can use data files from specialized applications such as Access, Paradox, or dBASE.
- *Styles, templates, libraries*—Only the more advanced word-processing programs enable you to create styles (special combinations of fonts, formatting, and text defined by you), templates (special document types that you often use, in which the style, format, and content must be the same), and libraries (special collections of styles and templates that you often use). Advanced tools like these enable you to define formatting elements (such as these bulleted list items with italicized lead-in) once instead of every time you use them. Tools like these also enable you to change special format or styles globally—once, instead of in every instance.
- *Conversion and filtering software*—Check the range of filters offered that enable you to convert files from other software. Check also to see how good a job the software does with the conversion (there may be so much touch-up work to do that you might as well start from scratch). Make sure that your software can convert documents to ASCII (strip out all the proprietary codes and dump out the straight text into a universally readable file).

Can Your Word-Processing Software Do This? *(continued)*

- *Document assembly*—Advanced word-processing software enables you to put together individual files and create a whole book, complete with automated tables of contents and indexes, cross-references, and other such book-level elements. When you're writing a book (or a book-length report), you may have dozens of separate files; document assembly takes all the pieces and puts them together for you.
- *Technical support*—Find out what kind of help is available from the manufacturer of the writing software you're considering. Do you have to pay a subscription fee in order to call with questions? Is there a 1-800 number? A 1-900 number?
- *Internet bulletin board*—Find out if there are Internet newsgroups or electronic mailing lists dedicated to the writing software you're considering. If calling the manufacturer directly is too expensive or too difficult (the lines may be continuously busy), the Internet alternative may be preferable.

TIPS ON WRITING WITH COMPUTERS

Pick Your Writing Software and Plan Your Document Carefully. A big problem in writing with a computer professionally is working with other professionals who use different writing software. Before you dive into a writing project, try to foresee how your document will be used and by whom. Will it need to be merged with other documents? If so, what software will be used? How much formatting will be necessary to convert to the software that is official for the project? If your document contains lots of intricate formatting, tables, charts, and so on, and if you're the one who has to convert to other software, you're in for some long tedious hours.

Consolidate All of Your Writing at the Computer. When people are just getting started writing with computers, they do much of the prewriting and planning offline—in other words, on paper. Some people even go so far as to write the rough draft on paper, saving the computer until time for the final copy. Why not just do it all at the computer? Work up the rough draft, take notes, create the outline, write the rough draft, edit and revise, finetune the formatting, and print out the final copy—all at the computer.

Rough-draft at the Computer. Some writers have a tendency to type at the computer keyboard as if they were producing the final copy (the writing equivalent of walking on eggshells). The beauty of writing at the computer is the ease of revision that it affords. There is no reason not to cut loose! And in general, regardless of the writing medium or instrument you use, the fewer inhibitions and restraints you put upon yourself, the better your rough drafting will go and the better your final output will be.

Save Often. As you work along on a document, save it every so often. If you do something really horrible to the document (such as a global formatting change that wrecks everything), you can revert to the last saved version. Most advanced software has an automatic save function built in; you can turn it on or off or adjust how often it automatically saves your work.

Make Backup Copies. How many times have you heard it said—make frequent backups! All that means is copying your files onto some other media, usually a diskette. Why? The main reason is that all hard files ultimately wear out or break down—some sooner than others. In any case, it's a dice roll exactly when it will happen. For example, in writing this book, we back up our files *after every single writing session.* If you use a laptop or notebook computer, be particularly careful to back up.

Automate Your Backup Process. You can make backing up easy and almost automatic. For example, if you have your important files all stored in one directory, you can set up a batch file (in DOS, those little files ending with **.BAT**) to handle the task. If it's a lot of data that won't fit on a disk, you can use a zip program to compress it all into one file that will fit on a diskette. You can set up an icon somewhere in your windowing program so that all you have to do is click on it to start the backup process. You can have the backup occur automatically every time you shut down the machine, or you can schedule backups to occur at certain times or on certain scheduled dates.

Be Careful with Multiple Copies. Watch out for the number of copies of a document you are working with. It gets absurdly easy to make different changes to different versions of the same document, and all of a sudden you have to consolidate all of those changes into one file.

Consolidate Your Personal Files in Special Subdirectories. Don't mix your own work with the files that are part of your writing software. Writing software usually has dozens if not hundreds of files that enable it to function. You don't want your own personal files getting lost in that. Instead, create and use a special directory for your own files. When it comes time to upgrade or migrate, you'll appreciate knowing what's yours and what's theirs!

Be Careful with Those Global Changes. The global search-and-replace function, for all the time it can save you, can wreak great havoc. An automatic search and replace will change *all* instances of the word or phrase it finds—not just the ones you had in mind. A prompted search and replace is much safer—slower, but far safer; it asks you each time whether to make the change.

Check Special Formatting and Do General Review on Hardcopy. Whenever you set up some special format in a document, such as a two-column list with bold column headings, print out the page on which it occurs to see how it looks. Often, a carefully designed format looks great on the screen but not so when printed out. And, in general, review your written documents on hardcopy to get a feel for the overall design and the general structure of the content. These "big-picture" issues are harder to perceive when you view a document strictly through a computer display.

Proofread on Hardcopy. Don't trust yourself to do a perfect proofreading job by staring at the computer screen. When you first start writing with computers, you may find that you do a lot of unnecessary printing out of your documents. It's a good thing to print out as little as possible—good for the trees, good for your printer supplies and hardware, and good for your finances. However, when you are finishing a document, you must print the thing out, go find some new location (a quiet spot in the company cafeteria midafternoon?), and proofread it carefully—the old-fashioned way—with a real pencil and real paper.

Be Careful with Fancy Typography. Avoid going crazy with fancy type styles and sizes. When we first get access to dozens of different fonts, different faces (bold, italics), and different sizes (6, 8, 9, 10, 11, . . . , 48 point sizes), our tendency is to have fun and use it all. However, the result is not fun for readers—it's garish and distracting. Overdoing it with the typographical effects is like dressing up in paisleys, polkadots, stripes, checks, and hounds-tooth designs. Figure 10-1 shows before-and-after versions of a page with severe "fontitis." Two points to remember:

- Typographical effects like bold or italics or typesize changes are for *emphasis*. However, if everything is bold or italics (or even both), then there is no emphasis.
- Be fairly conservative about page design. Readers are used to the printed word on the page looking a certain way. There's some variability, but if you get outside it then people begin to lose confidence and patience. Base what you do on well-established practice.

Use but Don't Trust Those Spell Checkers—Hire a Human. More and more, advanced word-processing and desktop publishing software packages not only

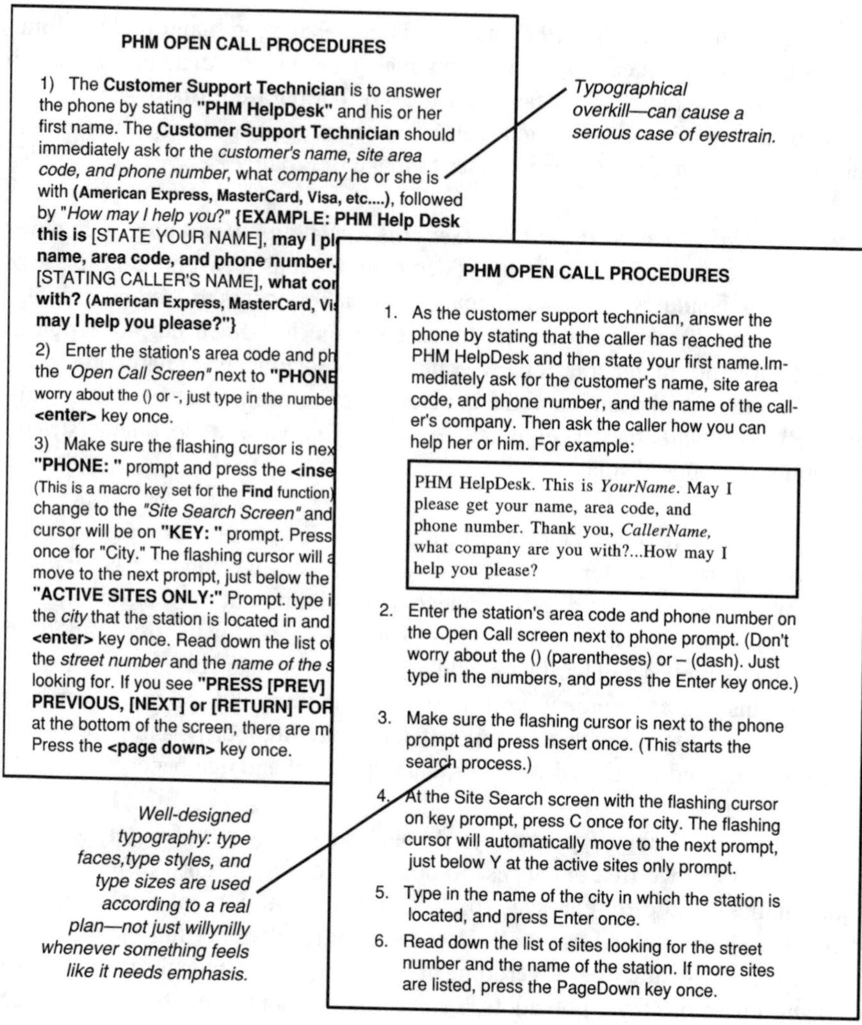

PHM OPEN CALL PROCEDURES

1) The **Customer Support Technician** is to answer the phone by stating **"PHM HelpDesk"** and his or her first name. The **Customer Support Technician** should immediately ask for the *customer's name, site area code, and phone number*, what *company* he or she is with **(American Express, MasterCard, Visa, etc....)**, followed by "*How may I help you?*" **{EXAMPLE: PHM Help Desk this is** [STATE YOUR NAME], **may I pl**
name, area code, and phone number.
[STATING CALLER'S NAME], **what co**
with? (American Express, MasterCard, Vi
may I help you please?"}

2) Enter the station's area code and ph
the *"Open Call Screen"* next to **"PHONE**
worry about the () or -, just type in the numbe
<enter> key once.

3) Make sure the flashing cursor is ne
"PHONE: " prompt and press the **<inse**
(This is a macro key set for the **Find** function)
change to the *"Site Search Screen"* and
cursor will be on **"KEY: "** prompt. Press
once for "City." The flashing cursor will a
move to the next prompt, just below the
"ACTIVE SITES ONLY:" Prompt. type i
the *city* that the station is located in and
<enter> key once. Read down the list o
the *street number* and the *name of the s*
looking for. If you see **"PRESS [PREV]**
PREVIOUS, [NEXT] or [RETURN] FOR
at the bottom of the screen, there are m
Press the **<page down>** key once.

Typographical overkill—can cause a serious case of eyestrain.

PHM OPEN CALL PROCEDURES

1. As the customer support technician, answer the phone by stating that the caller has reached the PHM HelpDesk and then state your first name. Immediately ask for the customer's name, site area code, and phone number, and the name of the caller's company. Then ask the caller how you can help her or him. For example:

   > PHM HelpDesk. This is *YourName*. May I please get your name, area code, and phone number. Thank you, *CallerName*, what company are you with?...How may I help you please?

2. Enter the station's area code and phone number on the Open Call screen next to phone prompt. (Don't worry about the () (parentheses) or – (dash). Just type in the numbers, and press the Enter key once.)
3. Make sure the flashing cursor is next to the phone prompt and press Insert once. (This starts the search process.)
4. At the Site Search screen with the flashing cursor on key prompt, press C once for city. The flashing cursor will automatically move to the next prompt, just below Y at the active sites only prompt.
5. Type in the name of the city in which the station is located, and press Enter once.
6. Read down the list of sites looking for the street number and the name of the station. If more sites are listed, press the PageDown key once.

Well-designed typography: type faces, type styles, and type sizes are used according to a real plan—not just willynilly whenever something feels like it needs emphasis.

Figure 10-1 A typographically gaudy file. It's hard not to use all that word-processing power at your fingertips. The most common problem is to go wild with bold, italics, bold italics, different fonts, and different type sizes. Remember, while it's fun to do, it's miserable to read.

have spell checkers but grammar checkers as well. Spell checkers are very useful—always take the time to run spell check on your documents. However, remember that spell-checking software can't help you with similar-sounding words such as *to, two, too,* or *where* and *wear,* or *your* and *you're,* or *principle* and *principal,* or *its* and *it's.*

Don't Waste Time on Grammar Checkers. Grammar checkers are quite another story. As of this writing, they are next to useless. The problem is that currently they are only 70% accurate, both in terms of what they miss and in terms of what they identify as wrong. That means that about one-third of your grammar errors are likely to just sit there quietly in your document unnoticed and that about one-third of the errors that are spotted are not errors at all. If you are concerned about these aspects of your professional writing, the best thing to do is to hire a professional editor. It's like fixing your plumbing—yes, you could probably do it yourself, but for triple the expense and time, and who knows how much extra grief. (If you're a student in a writing course, the idea is to learn—to go ahead and make the mistakes. The whole point is not so much to produce a perfect final product but to learn the techniques and processes of producing that final product.)

Proofread Last-minute Changes Extremely Carefully. In any writing or publishing project that has any importance at all, there are bound to be last-minute changes. On your "final" review, when you expect to make absolutely no further changes (and that's final!), you discover problems, or your partner rushes in with last-minute changes. Invariably, it is these changes inserted after the careful formal proofreading phase—these changes that are inserted hastily and with much tension hanging in the air—that contain glaring typos. There may not be time to print out and proofread again. If you can't do anything else, once you've input the changes, sit there and stare at them, reading and rereading them, daring them to turn into typos.

Be Careful with Automated Tables of Contents and Indexes. If you are doing a big writing project that involves a table of contents, an index, or both, you probably want to use advanced writing or publishing software that "automates" these and similar elements. You set up your text so that you can "run" the table of contents and index; the page numbers are then automatically generated. Problems arise, however, when you make subsequent changes to your text which in turn change the page numbers. That can make some of the page numbers in the TOC and index wrong. You have to regenerate the TOC and the index.

Never Trust a Pretty Page. One of the phenomena associated with powerful word-processing software is the awe of the well-crafted page. There it is: It's got fancy 6-pica indented text, hanging heads at decreasing 18-, 14-, and 10-point type sizes, italics and bold popping in the running text, nifty little running footers and headers alternating on recto and verso pages. It's tastefully designed, it's sophisticated, it's impressive—it's slick! But behind all that pretty typography may lurk a very rough draft! It's easy to be in awe of the pretty page and to assume that the same level of quality is manifest in the words themselves. That's another reason why many writers revert to relatively simple writing software. They don't get distracted by the surface glamor of what they are producing. They can get down and do the hard work on those words, sentences, and paragraphs—they can make sure those ideas are expressed well and are flowing smoothly.

Learn and Use Macros and Other Time-saving Devices. Invest the bit of time needed to learn how to create macros with your writing software. For example, imagine that you're writing about a corporation, product, or technology with a very long name. While there are several ways to do this, you can set up a macro that will automatically type the whole name out at a simple keystroke. Do you spend a lot of time formatting numbered or bulleted lists or special headings? Create a macro to handle that task. Or better yet, use advanced word-processing software's ability to set up styles, templates, and libraries.

GRAPHICS APPLICATIONS

For engineers, graphics software is likely to be as important as writing software, if not more so.[1] In your college courses, you probably were or will be introduced to one of the widely used graphics software packages, such as AutoCAD, CorelDraw, Adobe Illustrator, or Micrografx Designer. As with producing professional-looking text, producing professional-looking graphics has become increasingly the job of engineers as opposed to graphic artists. Just as you have to type and produce your own documents, you may have to do your own graphics. You can't expect to have secretaries or graphic artists always at your disposal.

Anybody who's ever touched a computer is probably familiar with MacPaint or Paintbrush. These are crayons compared to the sophisticated graphics packages available today such as those mentioned earlier. Advanced writing software and desktop publishing software also come with graphics tools. But if you have serious graphics needs, you'll probably need a full-featured package such as AutoCAD, CorelDraw, Adobe Illustrator, or Micrografx Designer.

As an engineer, look especially for features such as the following in the graphics software package you choose, and take the time necessary to learn how to use these features:

- *Portability*—Make sure that the graphics software you use can export and import graphics files to and from a reasonable variety of other graphics software. Much of your professional work is likely to be collaborative; you don't

[1] Our special thanks to Jim Reidy of IBM Corporation (Austin, Texas) for his contributions to this section on graphics application tools for engineers.

want to be prevented from sharing graphics with other colleagues on a project.

- *Clip-art repository*—Check the breadth and variety of clip art that comes with the software you choose. Think of the common symbols and objects you need in your work, and check to see that they are included in the graphics package. You will also want to be able to add to the clip-art repository with your own customized symbols.
- *Art-editing capabilities*—The graphics package you choose should be able to perform any editing feat that you can think of (rotating, stretching, shearing, mirror-imaging, cutting, pasting, adding fill patterns and color, and sizing as well as controlling line thicknesses and styles).
- *Good text capabilities*:
  - You should be able to size a graphic without changing the size of text associated with that graphic. For example, if you have 10-point text in a graphic, it should stay 10 point even if you double the size of that graphic.
  - You should also be able to convert text objects into graphic objects. Once you've transformed text this way, you can then transform it by angle, size, shape, and contour. For example, if you want to stretch a piece of text and set it at a 37-degree angle (or even curve it), most graphics software requires you first to convert it to a graphic object before you can do all these neat graphical tricks.
- *Zoom*—A good zoom feature is a must when you are working on intricate drawings; it will save your eyeballs!
- *Gravity (line-snap) and align capabilities*—You need this feature to achieve good sharp line intersections and alignments. Gravity is invaluable when you must close a shape that you've created with numerous lines so that you can fill it with a pattern or color.
- *Screen capture*—There are likely to be instances in your graphics work when you need to capture some portion of what is displaying on your computer screen. Screen capture creates a raster image file which you can then edit on a pixel-by-pixel basis.
- *Color capabilities*—Check to see that your graphics software is compatible with the Pantone Matching System (PMS) used by the printing industry. In addition, your software should be able to do color separation for you. If you use multicolored images, this color-separation feature will save your commercial printer time and save you money.
- *Printing graphics*—You should be able to print a graphic without having to open the file first. Many technical drawings have tens of thousands of lines in them; some have fill patterns (areas where objects are filled in by either color or some sort of grayscale) which add to the density of the art. Characteristics

like these can cause you to lose valuable time just waiting for graphics simply to open. On older machines, opening graphics can take up to an hour.

- *Object selecting and grouping*—Your software should give you a number of ways to select, reselect, or deselect one or more objects within a drawing; and you should be able to group and ungroup objects within a drawing.
- *Layering and overlays*—You should be able to move text or graphic objects within a drawing to the background or to the foreground, creating an overlay effect. And you should be able to choose whether those overlays are transparent or opaque.
- *Default customization*—Graphics software should enable you to set up your own defaults prior to starting an illustration. These defaults would control such things as line thickness, text size, colors, or grayscale.
- *Charts, graphs, and tables*—Although powerful methods for creating charts, graphs, and tables are becoming common in advanced word-processing software, you should expect these features in your graphics software as well. And they are likely to be more powerful and flexible in graphics software anyway. For charts and graphs, you simply enter your numerical data (or import or link it from some other application), and let the software generate your line graph, pie chart, or bar chart for you.

MISCELLANEOUS COMPUTER-BASED WRITING TOOLS

If you do a lot of writing on the computer, if you interact with colleagues over the Internet, there are several other tools you should know about and be comfortable using. Specifically, file-compression, antivirus, and scanning software (along with the scanning hardware) are likely to be very useful to you as a professional engineer.

Be aware, however, that most software programs such as IBM or Microsoft DOS as well as IBM OS/2 and Microsoft Windows are now including utilities such as file-compression and antivirus software. Typically, these programs are not as powerful or reliable as software separately available from the producers for whom these are specialties.

COMPRESSION SOFTWARE

Compression software is useful if you do lots of writing or graphics work with a computer. This type of software cuts the size of files to around one-half their original size. These compressed files are sometimes called "archive" or "zip" files. Imagine you're writing a report that takes up 1.8 MB on your hard drive. You can compress it to 0.9 MB or better and store it on a floppy disk, freeing up room on your computer.

Not only can compression software free up valuable hard disk space, it can help you in many other ways. It makes backing up your data much easier: In compressed form, backups take fewer diskettes. Compression software also makes sending data much easier: If you must send a big report to a colleague over the Internet, it's faster and cheaper to send the compressed version. This software also makes handling lots of files easier; you just compress them all into a single file. And of course you need compression software if you do any business on the Internet at all; most data on the net is in compressed form, and colleagues are likely to send you compressed data as well.

Make sure that your compression software enables you to create self-extracting files. When you have a compressed file that is self-extracting, all you do is type the name of that file as if it were a command, press Enter, and it uncompresses itself. You don't have to use separate decompression software to unpack it. This helps when your colleague or client doesn't know how to decompress files or doesn't have decompression software.

You can find compression software in most computer stores, or you can download it from the Internet and pay a registration fee. Also, check your operating system; newer versions are incorporating this type of software.

ANTIVIRUS SOFTWARE

You've probably heard of computer viruses by now—perhaps you've even been attacked by one. They disrupt your computer's software and ruin your data and programs. They can hide out in your system for months, even years, before they spring to life and erase your entire hard drive or do other sorts of damage. Also, they can hop onto diskettes and then infect any other computer they come in contact with. That's why you need another key tool—antivirus software—if you intend to use computers in your professional work. Unfortunately, most of us have to learn about this software by getting infected, losing data, spending hours eradicating the virus, and then installing antivirus software.

Antivirus software does several essential things for you; it can

- detect whether your computer is infected with a virus
- identify the virus in your computer
- eradicate the virus from your computer
- repair damage done by the virus (sometimes)
- act as a shield against incoming viruses in the future

Of course, virus programmers are out there constantly writing new virus programs; therefore, your antivirus software must constantly be updated. You do this by subscribing for updates with the antivirus software you purchase. You can download these updates from bulletin boards. Some antivirus programs build in an automatic download (which means of course you must have a modem and communications software).

As with file-compression software, antivirus software is now commonly being bundled with operating systems and with file-management utilities. Before you rush out and purchase antivirus software, check to see if your system already has it. Even so, the antivirus software that is bundled with other software such as operating systems or utilities may not be as powerful as that of the leading antivirus software producers; after all, it's their specialty. And in any case, remember that you can't just install an antivirus program and be done with it. Consider it as a subscription or a service (like a yearly termite inspection).

SCANNING EQUIPMENT

Another of the computer-based tools you'll find useful as an engineer is scanning hardware and software. Scanners, as you probably know, copy text, graphics, or both from print media into electronic media. For example, you could scan a diagram from a report into a computer file and then import it into a graphics software program or into your own report. (Of course, you're still obligated to cite your source for the graphic just as much as if you had borrowed text.)

When you go shopping for a scanner, check for these features:

- *Resolution.* The clarity, sharpness, detail, and resolution of the image that a scanner produces is, in part, a factor of its dots-per-square-inch (dpi) value. The more dpi's, the better the resolution. Until recently, 300 dpi was considered the norm; 600 is now considered standard.
- *Monochrome, grayscale, or color.* Early scanners did only line art and produced two colors, black and white. Grayscale scanners scan in varying shades of gray. Color scanners can scan 256 colors or over 16 million. A monochrome, or grayscale scanner is likely to meet all your professional engineering needs, and cost half as much as a color scanner.
- *Flat-bed, handheld, and pen scanners.* Flat-bed scanners look like smallish photocopying machines; handheld scanners look like a computer mouse or a wand; pen scanners are like a pen that you write with. Flat-bed scanners have traditionally been more accurate, produced higher resolution, handled much higher volume, and caused less technical trouble; they are also more expensive.
- *Page size or scanning area.* Flat-bed scanners start at the size of a legal sheet of paper; more expensive models have larger scanning areas. Handheld scanners have varying sizes of scanning area.

- *Paper-handling devices.* Some flat-bed scanners include various types of automatic sheet feeders. If you do high volumes of scanning, these are a big help.
- *Integration with FAX, photocopying, and printing functions.* Some machines combine the functions of a printer, FAX machine, photocopying machine, and scanner. If each of these functions in a particular machine is strong in its own right, then this is an option to consider.

Most scanners are packaged with some or all of the software you need to run the scanner and work with the scanned images:

- Actual scanning software that handles the details of manipulating the scanner and the computer and scanning and capturing the image.
- Graphics-manipulation and -conversion software that can be used to "crop" the image (prune it down to just the area you want), enlarge or reduce the image, change graphics values (such as color, intensity, hue), convert it to various standard graphic formats (such as .GIF, .PCX, or .TIFF), as well as other sorts of changes.
- Optical character recognition (OCR) software that is used to scan and convert text into text files that you can edit with your writing software. (Sometimes this software component is not included in the scanner package.)

EXERCISES

Using a common advanced word-processing or desktop-publishing software package you are comfortable with, do any of the following that are unfamiliar:

1. Open a document and create a macro that handles a repetitive task such as finding a certain text string, modifying it, italicizing it, and then putting it into bulleted-list format. Once you've successfully created the macro, bind it to a key sequence (for example, Alt-A), and then store it for future use. (Exit the software, then return, and see if you can use the macro again.)
2. Create a "style," for example, a specially designed heading, bulleted-list item, or indented text with an unusual font. Save it under a name you define for future use. (Exit the software, then return, and see if you can use that style.)
3. Create a document template, for example, for the business-letter format that a hypothetical engineering firm might require its staff to use. Give it a special name that you define, and then store it for future use. (Exit the software, then return, and see if you can use the template.)
4. Set up a simple document and accompanying data file with which you can test the "merge" function of your software. For example, set up a "shell" business letter or

report with variables for changeable information such as company, product, or addressee names. Set up the corresponding data file with the variable information, and merge that data into the "shell" document.

5. Interview at least three professional engineers concerning their uses of computers:
   a. Which advanced word-processing software (for example, Word, WordPerfect, Ami-Pro) or desktop-publishing software (for example, Interleaf, FrameMaker, QuarkXPress) does the firm use?
   b. Which graphics programs (for example, AutoCAD, CorelDraw, Freelance, Adobe Illustrator) do the members of the firm use?
   c. How competent are staff engineers expected to be with these software programs? Is in-house technical support available?

BIBLIOGRAPHY

Few books survey and evaluate whole areas of software such as word-processing, communications, or graphics. To get some background in computer products, scan the tables of contents of a year's worth of magazines like the ones listed below. Typically, these magazines have articles that provide detailed comparisons of computer products such as scanners, modems, CD-ROMs, all sorts of software, and, of course, computers. Occasionally, these magazines will have short articles on technical aspects of these products, for example, how a virus works. At the turn of the year, these magazines typically present their pick of the best products for the year.

Computer Shopper. A publication of Ziff-Davis Publishing Company providing many in-depth reviews of products and technical articles.

Macworld. A monthly publication of Macworld Communications, Inc., on Macintosh user interests.

Macuser. A monthly publication of Ziff-Davis Publishing Company on Macintosh user interests.

PC Computing. A semimonthly publication of Ziff-Davis Publishing Company on PC user interests.

PC Magazine. A monthly publication of Ziff-Davis Publishing Company on PC user interests.

PC World. A monthly publication of Ziff-Davis Publishing Company on PC user interests.

INDEX

Z